Salters-Nuffield Advanced B

AS level

Student book 1

Trial Edition

This trial edition may contain typographical errors and other small faults because it is a first proof of the material. We hope this will not detract from reading it. Errors will be rectified in the final published version.

Heinemann Educational Publishers
Halley Court, Jordan Hill, Oxford, OX2 8EJ
a division of Reed Educational & Professional Publishing Ltd
Heinemann is a registered trademark of Reed Educational & Professional Publishing Ltd

OXFORD MELBOURNE AUCKLAND
JOHANNESBURG BLANTYRE GABORONE
IBADAN PORTSMOUTH NH (USA) CHICAGO

First published 2002

ISBN 0 435 62834 8

05 04 03 02 01

10 9 8 7 6 5 4 3 2 1

Designed and typeset by J&L Composition Ltd, Filey

Printed and bound in Great Britain by The Bath Press Ltd, Bath

Acknowledgements

The authors and publishers would like to thank the following for permission to use photographs:

p3 Getty Images; p5 John Birdsall Photography; p6 and 7 Mark Tolley; p10 SPL/John Radcliffe Hospital; p11 SPL/Biophoto; p13 L-R: SPL, SPL/CNRI, SPL/Biophoto associates; p16 T-B: SPL/Professor P.M. Motta/G. Macciarelli/S.a. Nottola, SPL/CNRI; p17 British Heart Foundation; p18 SPL/Deep Light Productions; p22 SPL/Dr P. Marazzi; T-B: p28 CEC/Marc N Boulton; p31 Dairy Crest; p38 Fiennes/Stoud/Howell; p41 CEC/Mark N Boulton; p50 SPL/Sheila Terry; p51 CEC/ Mark N Boulton; p54 L-R: Mother and Baby Picture Lirary/Angela Spain, Mother and Baby Pictuer Library/EMAP; p56 SPL/Jurgen Berger, Max-Planck Institute; p58 SPL/BSIP; p62 T: SPL/P. Motta/Dept of Anatomy/University "La Sapienza" Rome, B: SPL/Patricia Chulz, Pete Arnold Inc; p70 SPL/Photo Researchers; p80 SPL/Eye of Science; p88 SPL/John Bavosi; p92 T: SPL/Saturn Stills, B: SPL/Pascal Goetgheluck.
(T=Top, B=Bottom, L=Left, R=Right. SPL = Science Photo Library)

The publishers have made every effort to trace the copyright holders, but if they have inadvertently overlooked any, they will be pleased to make the necessary arrangements at the first opportunity.
Picture research by Liz Savery

Unit 1

Contributors

Many people from schools, colleges, universities, industries and the professions have contributed to the Salters-Nuffield Advanced Biology project. They include the following.

Central team

Angela Hall, Nuffield Curriculum Projects Centre
Michael Reiss (Director), Institute of Education, University of London
Anne Scott, University of York Science Education Group
Sarah Codrington, Nuffield Curriculum Projects Centre
Nancy Newton (Secretary), University of York Science Education Group

Advisory Committee

Professor R McNeill Alexander FRS	University of Leeds
Dr Allan Baxter	GlaxoSmithKline
Professor Sir Tom Blundell FRS (Chair)	University of Cambridge
Professor Kay Davies CBE	University of Oxford
Professor Sir John Krebs FRS	Food Standards Agency
Professor John Lawton FRS	Natural Environment Research Council
Professor Peter Lillford CBE	University of York
Dr Roger Lock	University of Birmingham
Professor Angela McFarlane	University of Bristol
Dr Alan Munro	University of Cambridge
Professor Lord Winston	Imperial College, London

Unit 1 authors

Topic 1

Glen Balmer	Watford Grammar School
Alan Clamp	Ealing Tutorial College
Ginny Hales	Cambridge Regional College
Gill Hickman	Ringwood School
Liz Hodgson (Team Leader)	Greenhead College, Huddersfield

Topic 2

Jon Duveen (Team Leader)	City & Islington College, London
Brian Ford	The Sixth Form College, Colchester
Steve Hall	King Edward VI School, Southampton
Pauline Lowrie	Sir John Deane's College, Northwich
Jamie Shackleton	Cambridge Regional College

Other materials

Malcolm Ingram	
Rachel Hadi-Talab	Institute of Education, University of London
Jacquie Punter	Brighton, Hove and Sussex Sixth Form College
Cathy Rowell	

We would also like to thank the following for their advice and assistance.

John Holman	University of York
Andrew Hunt	Nuffield Curriculum Projects Centre
Jenny Lewis	University of Leeds

Sponsors

The Salters Institute
The Nuffield Foundation
The Wellcome Trust
ICI plc
Boots plc
Pfizer Limited
Zeneca Agrochemicals
The Royal Society of Chemistry

Contents

How to use this book

Welcome to unit one of the new Salters-Nuffield Advanced Biology course. This is the first of two student textbooks for the AS course.

This book covers two topics: Topic 1 "Lifestyle, health and risk" and Topic 2 "Genes and health". In each topic we begin with a context and then draw out the underlying biological concepts.

Within each topic you will develop your knowledge and understanding of these biological concepts. In later topics you will meet many of these again in new contexts. This will help you to further your understanding.

Each topic contains a number of features which we hope will help your learning.

Main text

The **text** presents the contexts of the unit and explains the relevant biology. Key terms are printed in **bold**. We advise that you check you understand the meanings of these words.

The main text also contains occasional boxes. As their name suggests, '**Key biological principle' boxes** highlight some fundamental biological principles. We also have '**Nice-to-know**' boxes. These contain material that you won't be examined on but we hope you will find interesting.

At the end of the main text for each topic you will find a **summary**. The summary contains a bullet-pointed list of what you need to have learnt from the topic for the unit exam.

Activities

Throughout the unit you will find references to activities. Activities include practicals, interactive on-line exercises, role-plays, issues for debate, etc. Details of all the activities can be accessed on-line and most can also be used in a paper form. Your teacher or lecturer will guide you as to which activities to do and when.

Each topic begins with a particular sort of activity called a GCSE review test. This enables you to check that you can remember and understand enough from GCSE to start the topic. Each topic finishes with an end-of-topic text. This allows you to see how much of the topic you have understood and remembered. Both the review test and the end-of-topic test can be completed interactively using the website.

Skills support

As part of the dedicated website you can find support on related science, maths and ICT skills.

Questions

Every now and again in the text we have provided questions. These are intended to get you to think about the material. Answers are provided at the end of the book. We'll leave it up to you to decide whether or not the best strategy is to think about the questions before you look at the answers. . . .

Any comments?

Finally, this is a pilot. We are therefore keen to hear from you with suggestions about how the course could be improved. You can e-mail us at contact@snab.org.uk

or write to us at

The Salters-Nuffield Advanced Biology Project, Science Education Group, University of York, Heslington, York YO10 5DD

Best of luck!

Lifestyle, health and risk

Lifestyle, health and risk

Why a topic called *Lifestyle, health and risk?*

Congratulations on making it this far; not everyone who started the journey has been as lucky. In the UK only about 50% of conceptions lead to live births and approximately 6 in every 1000 new-born babies do not survive their first year of life (Figure 1.1). After celebrating your first birthday there seem to be fewer dangers. Less than 2 in every 1000 children die between the ages of 1 and 14 years old. All in all, life *is* a risky business (Figure 1.2).

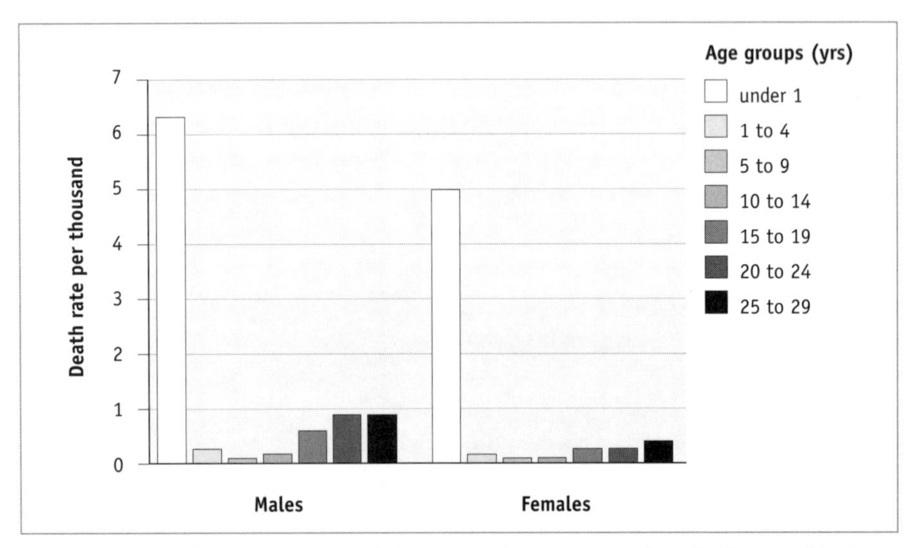

▲ **Figure 1.1** Death rates per 1000 population per year by age group and sex. Is life more risky for boys? (England and Wales Office for National Statistics 1998.)

In everything we do there is some risk. Normally we only think something is risky if there is the obvious potential for a harmful outcome. Snowboarding, parachute jumping and taking ecstasy are thought of as risky activities, but even crossing the road, jogging or sitting in the sun have risks, and many people take actions to reduce them.

Risks to health are often not so apparent as someone parachute jumping. Do people even realise they are at risk? Do we underestimate lifestyle risks?

What we eat and drink, and the activities we take part in, all affect our health and well-being. Every day we make choices that may have short and long-term consequences of which we may only be vaguely aware. What are the health risks we are subjecting ourselves to? Will a cooked breakfast set us up for the day or will it set us on course for heart disease? Does the 10-minute walk to work really make a difference to our health?

Cardiovascular disease is the biggest killer in the UK. Does everyone have the same risk? Can we assess and reduce the risk to our health? Do we need to? Is our perception of risk at odds with reality?

Overview of the biological principles covered in this topic

This topic will introduce the concept of risks to health. You will study the relative sizes of risks, how they are assessed and what affects our perception of risk. We will also look at how health risks may be affected by lifestyle choices.

Building on the knowledge of the circulatory system gained at GCSE you will study the heart and circulation and understand how these are affected by our choice of diet and activity.

You will look in some detail at the biochemistry of our food to help gain a detailed understanding of some of the current thinking among doctors and other scientists with regard to the choice of foods to reduce the risks to our health.

▲ **Figure 1.2** A 15–24 year old male in the UK is over three times as likely to have a fatal accident than is a female of the same age.

Review test

Are you ready to tackle Topic 1 – Lifestyle, health and risk

Complete the GCSE review before you start. **AS01RVT01**

1.1 What do we mean by risk?

Risk is defined as "the probability of occurrence of some event or outcome". Probability has a precise mathematical meaning and can be calculated to give a numerical value. Do not panic – the maths is simple.

Taking a risk is a bit like throwing a die (singular of 'dice'). The chance that you will have an accident or succumb to a disease (or throw a six) can be calculated. An accident or illness will not *necessarily* befall you, but by looking at past circumstances of people who have taken the same risk, the chance that you will suffer the same fate can be estimated to a reasonable degree of accuracy.

Working out probabilities

There are six faces on a standard die. Only one face has six dots; therefore the chance of throwing a six is 1 in 6 (provided the die is not loaded). This 1 in 6 chance is true for every throw of every dice, and for each of the six numbers on standard dice. Throwing a six does not affect the outcome of the next or any subsequent throws: providing nothing about the die is changed, the probability of throwing a six always remains 1 in 6. Scientists tend to express '1 in 6' as a decimal: 0.1666666 recurring. In other words, each time you throw a standard die, you have got about a 17% (= 0.17) chance of getting a 1, about a 17% chance of getting a 2, etc..

Using probability to estimate the risks to health

In 1998, 19 523 people in the UK died due to injuries or poisoning. The total UK population at the time was 59 236 500 so we can calculate the average risk of someone in the UK dying from injuries or poisoning as 19 523 in 59 236 500, or 1 in 3034, or 0.00033. Assuming the proportion of people that die from accidents or poisoning remains much the same each year this gives an estimate of the risk for any year.

How does this risk of death compare to others?

Q1.1 Look at the causes of death listed below and put them in order, from the most likely to the least. If you feel like it, have a go at estimating for each of these causes of death the probability of someone in the UK dying from it during the course of a year:

- **Accidental poisoning**
- **Heart disease**
- **Injury purposely inflicted by another person**
- **Lightning**
- **Lung cancer**
- **Railway accidents**
- **Road accidents.**

Did you get it right?

People frequently get it wrong, underestimating or overestimating risk. We can say that there is a about a one in 5000 risk of each of us dying from influenza in any one year, a one in 100 000 risk of us being murdered in the next twelve months and a one in ten million risk of us being hit by lightening in a year. However, recent work on risk has concentrated not so much on numbers such as these but on the **perception** of risk.

The significance of the perception of risk can be illustrated by a decision in September 2001 of the American Red Cross, which provides about half of the USA's blood supplies. They decided to ban all blood donations from anyone who has spent six months or more in any European country since 1980. The reason for this is because of the risk of transmitting variant Creutzfeldt-Jakob disease (vCJD) – the human form of bovine spongiform encephalopathy (BSE) – through blood transfusion. Experts agree that there is a *chance* of this happening. Yet there is to date not a single known case of this *actually* having happened. Indeed, as the USA is short of blood for blood transfusions it is very possible that more people will die as a result of this 'safety precaution' than would have had it not been introduced.

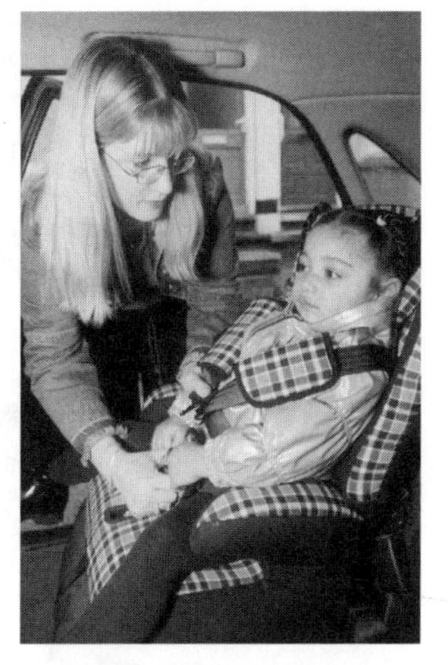

▲ **Figure 1.3** Some research suggests that young children who wear rear seat belts are *more* likely to die. Health risks are greatly affected by human behaviour.

So why has America banned European blood donations? Probably because of public perceptions of the risk of contracting vCJD. It is now known that the following circumstances make it likely that people will overestimate the risk of something happening:

- the risk is involuntary
- the risk is not natural
- the risk is unfamiliar
- the risk is dreaded
- the risk is unfair.

If you look at this list you should be able to see why people may greatly overestimate some risks (such as the chances of contracting vCJD from blood transfusions) while underestimating others (such as the dangers of driving a bit over the speed limit or playing on a frozen lake). A useful distinction is sometimes made between risk and uncertainty. When we lack the data to estimate a risk precisely, we are *uncertain* about the risk. For example, we are uncertain about the environmental consequences of many chemicals.

Activity

Activity 1.1 gets you to estimate risks for a range of diseases using the National Office for Statistics data on causes of death in the UK. **AS01ACT01**

Nowadays many risk experts argue that perceptions of risk are what really drive people's behaviours. Consider what happened when it became compulsory in the UK to use seat belts for children in the rear seats of cars (Figure 1.3). The number of children killed and injured *increased*. How could this be? John Adams, an academic at University College London, argues that this is because the parents driving the cars felt safer once their children were wearing seat belts and so drove slightly less carefully. Unfortunately, this change in their driving behaviour was more than enough to compensate for any extra protection provided by the seat belts.

There is a tendency to overestimate the risks of sudden imposed dangers where the consequences are severe, and underestimate a risk if it has an effect in the long-term future, even if that effect is severe.

Mark's Story

On 28th July 1995 something momentous happened that changed my life ...

▲ **Figure 1.4** Mark at 15.

I was sitting in my bedroom playing on my computer when I started to feel dizzy with a slight headache. Standing, I lost all balance and was feeling very poorly. I think I can remember trying to get downstairs and into the kitchen before fainting. People say that unconscious people can still hear. I don't know if it's true but I can remember my Dad phoning for a doctor and that was it. It took five minutes from me being an average 15-year-old to being in a coma.

I was rushed to Redditch Alexandra Hospital where they did some reaction tests on me. They asked my parents questions about my lifestyle (did I smoke, take drugs etc?). Failing to respond to any stimulus, I was transferred in an ambulance to Coventry Walsgrave Neurological Ward. Following CT and MRI scans on my brain it was concluded that I had suffered a stroke. My parents signed the consent form for me to have an operation lasting many hours. I was given about a 30% chance of survival.

They stopped the bleed by clipping the blood vessels that had burst with metal clips, removing the excess blood with a vacuum. I was then transferred to the intensive care unit to see if I would recover. Within a couple of days I was conscious and day-by-day regained my sight, hearing and movement (although walking and speech was still distorted.) They had shaved all my hair off!

I had a remarkably quick recovery considering the severity of the operation. I was talking again (although slurred and jumbled) within 5 days. By the end of the week, I was transferred back to Redditch Alexandra Hospital to continue the rest of my recovery.

There I received occupational therapy, physiotherapy and speech & language therapy to improve my co-ordination, speech and strength. Within seven days I could walk aided and talk better – I was then discharged to complete my recovery at home. I was given a wheelchair and was admitted for therapy as an outpatient. The occupational therapy trained my ability to perform everyday tasks. They made me make tea, do jigsaws etc. to improve my cognitive skills.

Another effect that the haemorrhage had on me was that the whole right-hand side of my body was weakened (the haemorrhage happened on the left side of my brain) and things that I took for granted before became a challenge. My left hand compensated for the weakness and gradually I became stronger, albeit on my former weaker side!

Three weeks later I returned to Coventry Walsgrave for an angiogram, where an X-ray dye was injected into my veins which showed up my blood vessels on a scan. However this showed that there was still a bleed occurring and so I was prepared for surgery once again.

The operation was lengthy, but not an emergency. However I was still warned of the dangers of such surgery. The operation did not leave me with much disability this time, and I woke up within a day of being transferred back into the intensive care unit again. Speech and movement were regained quickly. I was discharged to outpatient within 3 weeks, after undergoing another angiogram, and MRI and CT scans on my brain. Embarrassingly, they had shaven only *half* of my hair off this time!

▲ **Figure 1.5** The experience is not stopping Mark living life to the full.

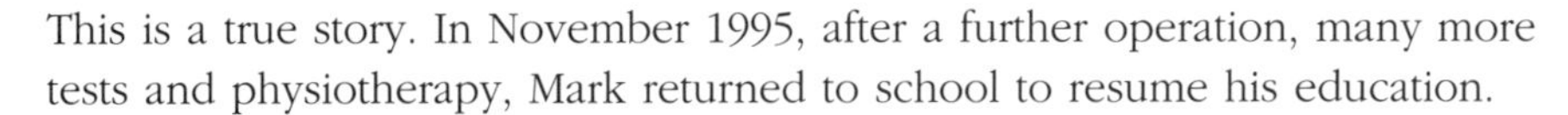

This is a true story. In November 1995, after a further operation, many more tests and physiotherapy, Mark returned to school to resume his education.

Mark had a **stroke**, one of the forms of cardiovascular disease. It is rare for someone as young as Mark to suffer a stroke. Why did it happen? Was he in a high risk group?

1.2 What is cardiovascular disease?

Cardiovascular diseases (CVDs) are diseases of the heart and circulation. They are the main cause of death in the UK (Figure 1.6), accounting for over

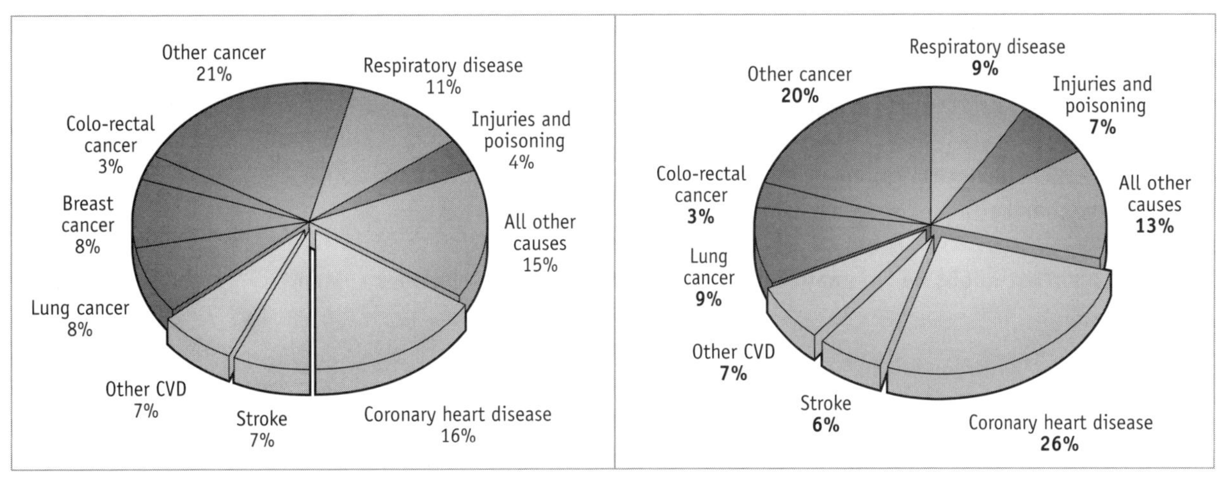

▲ **Figure 1.6** Death by cause in the UK in 1998 for females under the age of 75 years (left) and males under the age of 75 years (right). One person dies of heart disease every three minutes.

250000 deaths a year. More than one in three people in the UK die from cardiovascular diseases. The main forms of cardiovascular diseases are **coronary heart disease** (CHD) and **stroke**.

Key biological principle: why have a heart and circulation?

- In the absence of some sort of **mass flow**, substances can only be moved around within an organism by **diffusion**, i.e. the random movement of molecules.
- Diffusion is fast enough over very short distances to mean that very small organisms, such as unicellular creatures, can rely on it. Larger organisms need to move substances such as oxygen, carbon dioxide and the products of digestion over distances that are too great for diffusion to be enough.
- In many animal species, including ourselves, one or more **hearts** pump **blood** round the body. (The humble earthworm, for example, has five hearts.)
- Insects and some other animal groups have an **open circulatory system** in which the blood circulates in large open spaces.
- Large animals, including all vertebrates, have a **closed circulatory system** in which the blood is enclosed within tubes. Valves ensure that flow is only in one direction. The blood leaves the heart under pressure and flows along **arteries** and then **arterioles** (small arteries) to **capillaries**. There are extremely large numbers of capillaries and they come into close association with most of the cells in the body. After passing along the capillaries, the blood returns to the heart by means of **venules** (small veins) and **veins**.
- Animals with a closed circulatory system either have single circulation or double circulation. **Single circulation** is found, for example, in fish (Figure 1.7). The heart pumps deoxygenated blood to the gills. Here **gaseous exchange** takes place: there is a net diffusion of carbon dioxide from the blood into the water that surrounds the fish and a net diffusion of oxygen from this water into the blood. The blood leaving the gills then flows round the rest of the body before eventually returning to the heart. Note that the blood flows through the heart once for each complete circuit of the body.
- Birds and mammals, though, have **double circulation**, as shown in Figure 1.7. The heart (right ventricle) pumps deoxygenated blood to the lungs but then the blood returns to the heart to be pumped a second time (by the left ventricle) before going to the rest of the body. This means that the blood flows through the heart twice for each complete circuit of the body. The extra 'boost' that the heart is able to give to the blood returning from the lungs means that it takes the blood much less time to circulate round the whole body than would otherwise be the case. This allows birds and mammals to have a higher **metabolic rate**.

Q1.2 Why do only small animals have an open circulatory system?

Q1.3 What are the advantages of having a double circulation system?

Q1.4 Fish have two-chamber hearts and mammals have four-chamber hearts. Sketch what the three-chamber heart of an amphibian might look like.

Q1.5 What might be the major disadvantage of this three-chamber system?

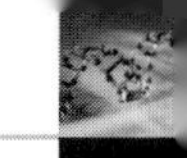

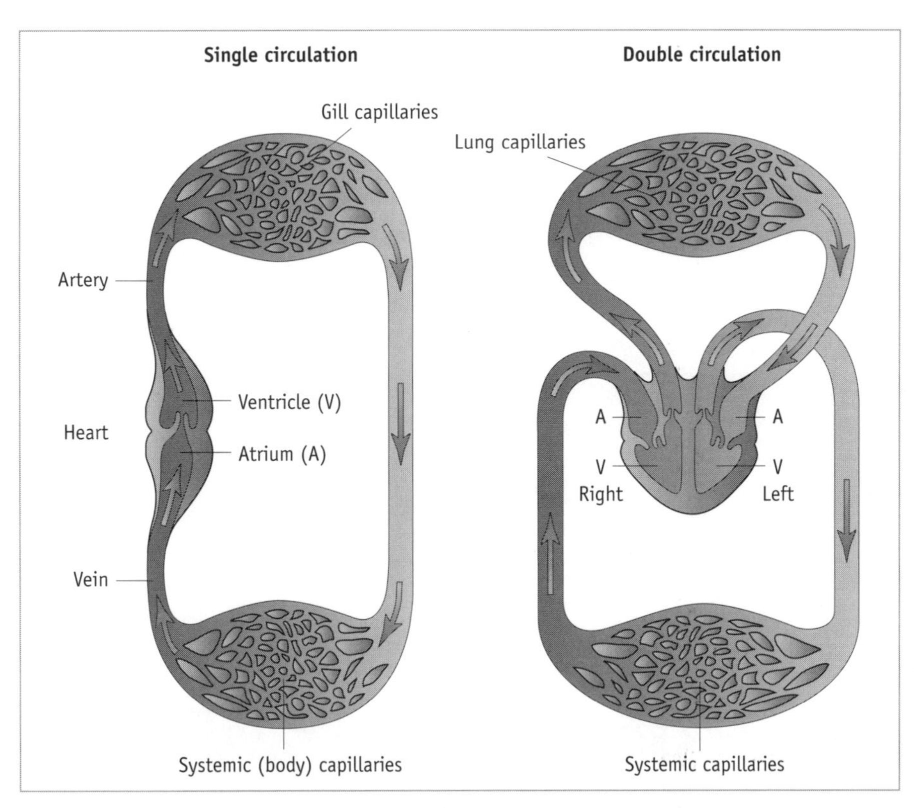

▲ **Figure 1.7** Fish have a single circulation. Birds and mammals have a double circulation.

About half of all deaths from cardiovascular diseases are from coronary heart disease and about a quarter are from stroke. Coronary heart disease is the most common cause of death in the UK. One in four men and one in five women die from the disease. To check out the most recent figures see the National Statistics Office web site. The URL is given in Activity 1.1.

What causes cardiovascular disease?

Atherosclerosis, or 'hardening of the arteries', is the name given to the disease process that leads to coronary heart disease and **ischaemic stroke**. 'Ischaemia' means 'insufficient supply of blood to part of the body'. So an ischaemic stroke results when not enough blood gets to the brain. In atherosclerosis fatty deposits narrow the arteries restricting blood flow. Eventually the deposits can either block an artery directly or increase the chance of it being blocked by a blood clot (**thrombosis**). The blood supply can be blocked completely and the tissue supplied by the artery is said to be ischaemic (without blood). If the ischaemia is prolonged, the affected cells are permanently damaged. In the arteries supplying the heart this results in a heart attack (**myocardial infarction**); in the arteries supplying the brain it results in a stroke.

Activity

Activities 1.2 and **1.3** let you look in detail at the structure of a mammalian heart using either a dissection or a simulation.
AS01ACT02 (actual dissection)
AS01ACT03 (simulated dissection)

What are the differences between arteries and veins?

Study Figure 1.8 and locate the arteries carrying blood away from the heart and the veins returning blood to the heart. Since the heart is a muscle it must constantly receive a fresh supply of blood. You might think that receiving a blood supply would never be a problem for the heart. However, the heart is unable to use any of the blood contained within its pumping chambers directly. Instead, the heart muscle is supplied with blood through two vessels called the **coronary arteries**.

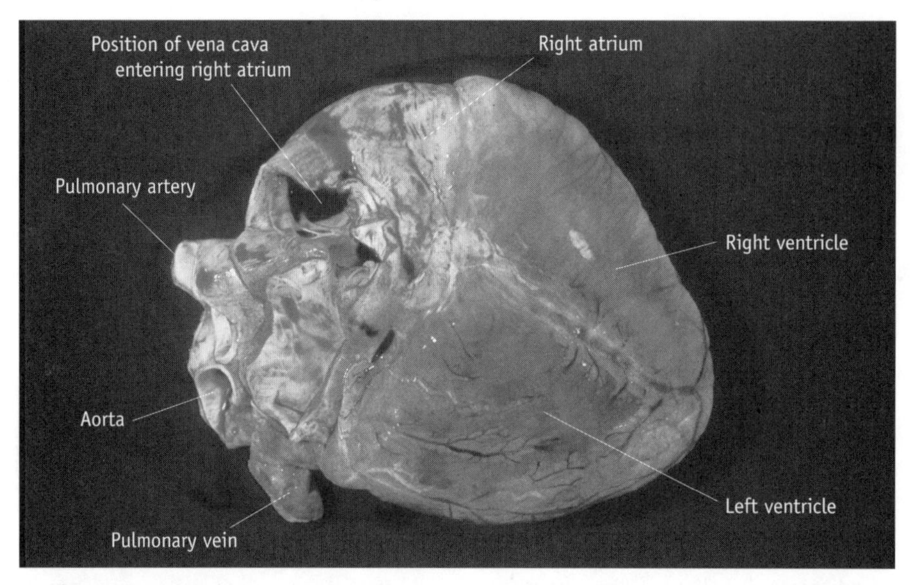

▲ **Figure 1.8** A normal human heart.

Activity

Activity 1.4 lets you investigate how the structure of blood vessels relates to their function. **AS01ACT04**

Every time the heart contracts (**systole**) blood is forced into arteries and the elastic walls stretch to accommodate the blood. During **diastole** (relaxation of the heart) the elasticity of the arteries causes their walls to contract behind the blood, pushing the blood forward. The blood moves along the length of the artery as each adjacent section is thus extended and recoils. This maintains a continual, periodic flow of blood through the arteries and capillaries. This **pulse** of blood moving through the artery can be felt anywhere an artery passes over a bone close to the skin. By the time the blood reaches the smaller arteries and capillaries there is a steady flow of blood. In the capillaries this allows exchange between blood and the surrounding cells through the one-cell thick capillary walls.

The heart has a less direct effect on the transport of blood through the veins. In the veins blood flow is assisted by the contraction of skeletal muscles during movement of limbs and breathing. The steady flow without pulses of

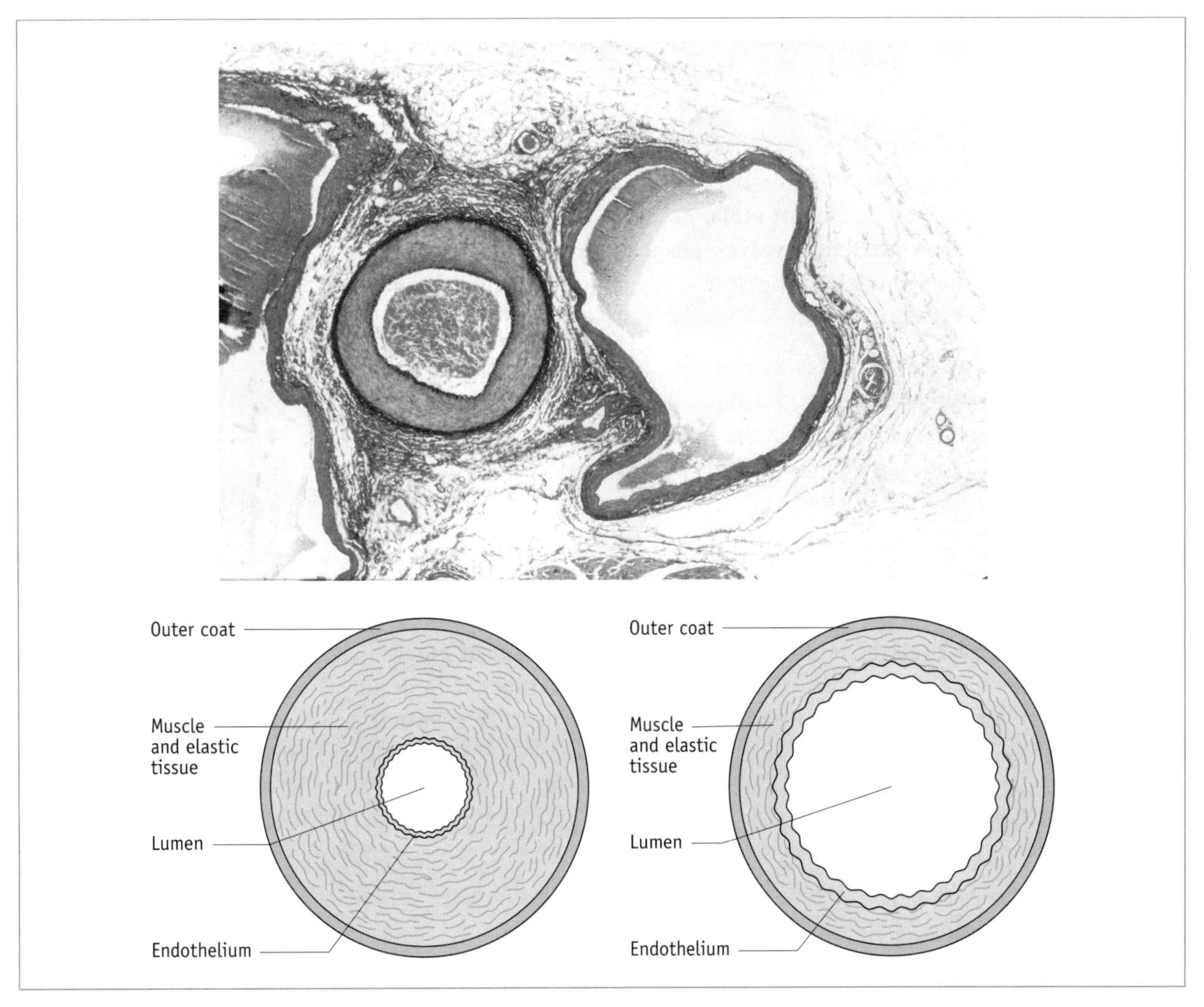

▲ **Figure 1.9** Identify three differences between an artery and a vein that can be observed on the photomicrographs and accompanying diagrams.

blood means that the blood is under low pressure. Because of this the walls of veins are much thinner than those of arteries.

Q1.6 List the features shown in Figure 1.9 that enable the artery to withstand high pressure and then recoil to maintain a steady flow of blood. Note: collagen is a tough fibrous protein.

How does the heart pump blood into the arteries?

Give a tennis ball a good, hard squeeze. You're using about the same amount of force that your heart uses in a single contraction to pump blood out to the body. Even at rest, the muscles of the heart work hard – weight for weight, harder than the leg muscles of a person running.

The chambers of the heart alternately contract (systole) and relax (diastole) in a rhythmic cycle. One complete sequence of filling and pumping blood is called a **cardiac cycle**, or heart beat. During systole, cardiac muscle contracts and the heart pumps blood out through the aorta and pulmonary arteries; during diastole, cardiac muscle relaxes and the heart fills with blood.

The cardiac cycle can be simplified into three phases: **atrial systole**, **ventricular systole** and **diastole**. The events that occur during each of the stages are shown in Figure 1.10.

Blood returns to the heart due to the action of skeletal and gaseous exchange (breathing) muscles as you move and breath. Blood under low pressure flows into the **left** and **right atria**. As the atria fill, the pressure of blood against the **atrio-ventricular valves** pushes them open and blood begins to leak into the **ventricles**. The atria walls contract forcing more blood into the ventricles, this is known as atrial systole.

This is immediately followed by ventricular systole. The ventricles contract from the base of the heart upwards, pushing blood up and out through the arteries. The pressure of blood against the atrio-ventricular valves closes them and prevents blood flowing backwards into the atria. The atria and ventricles then relax during diastole.

Closing of the atrio-ventricular valves and then the **semilunar valves** create the characteristic sounds of the heart.

Q1.7 When the heart relaxes in diastole how is the movement of blood from the arteries back into the ventricles, which might be thought to occur due to the action of gravity and the elastic recoil of the heart, prevented?

Activity

Activity 1.5 lets you test your knowledge of the cardiac cycle. **AS01ACT05**

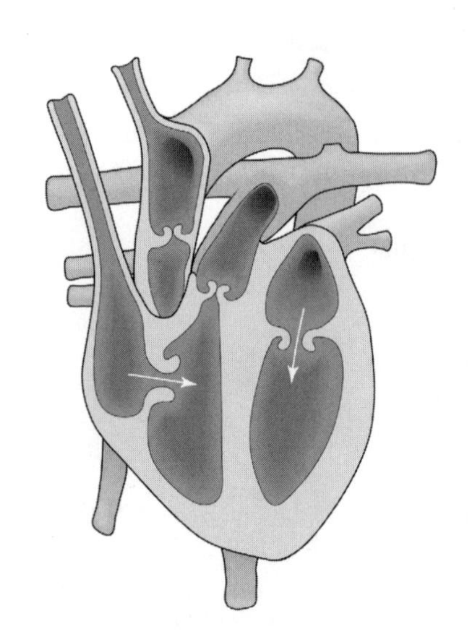

Atrial systole
The atria contract, forcing blood into the ventricles.

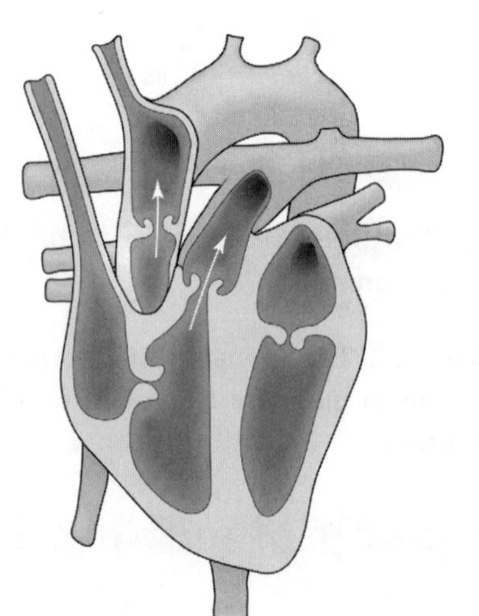

Ventricular systole
Contraction of the ventricles pushes blood up into the arteries.

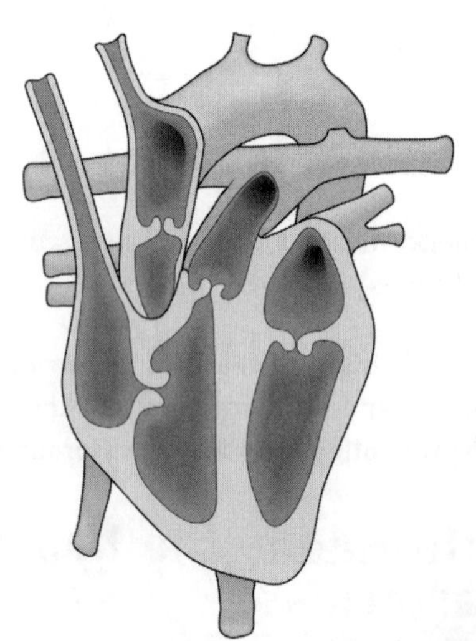

Diastole
Elastic recoil as the heart relaxes causes low pressure in heart, helping to re-fill the chambers with blood from the veins.

▲ **Figure 1.10** The three stages of the cardiac cycle.

What happens in atherosclerosis?

Atherosclerosis can be triggered by a number of factors. Here is the course of events whatever the trigger:

1. The delicate layer of cells that lines the inside of an artery (Figure 1.11a), separating the blood that flows along the artery from the muscular wall, becomes damaged for some reason. For instance, this damage can result from high blood pressure, which puts an extra strain on the layer of cells, or result from some of the toxins in cigarette smoke getting into the blood stream.
2. Once the inner lining of the artery is breached, there is an inflammatory response. Large white cells leave the blood vessel and move into the wall. These cells accumulate chemicals from the blood, particularly **cholesterol**. The technical name for the resulting deposit is an **atheroma**.
3. **Calcium salts** and **fibrous tissue** also build up at the site and a hard swelling called a **plaque** results on the inner wall of the artery. The build up of fibrous tissue means that the artery wall loses some of its elasticity: in other words, it hardens. The ancient Greek word for 'hardening' is 'sclerosis' and since medical terms are often based on Greek or Latin, we have the word 'atherosclerosis'.
4. Plaques cause the artery to become narrower (Figure 1.11b). This makes it more difficult for the heart to pump blood round the body and can lead to a rise in blood pressure. Now there is a dangerous **positive feedback** building up. Plaques lead to raised blood pressure and raised blood pressure makes it more likely that further plaques will form.

The individual, at this stage, is probably unaware of any problem, but if the arteries become very narrow or blocked then they cannot supply enough blood to meet the needs of the tissue for oxygen and nutrients. The tissue can no longer function normally and symptoms will become evident to the sufferer.

Why does the blood clot within the artery?

When blood vessel walls are damaged or blood flows very slowly the likelihood of a blood clot occurring is greatly increased (Figure 1.11c). **Platelets** (cell fragments without a nucleus) come into contact with the

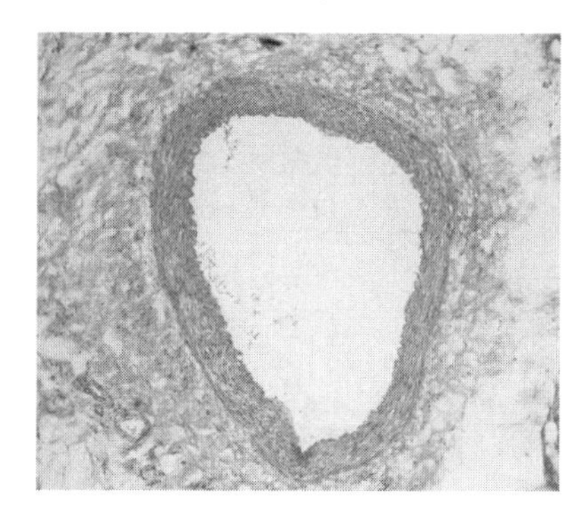

▲ **Figure 1.11a** Photomicrograph of a normal, healthy coronary artery showing no thickening of the arterial wall and the lumen is large.

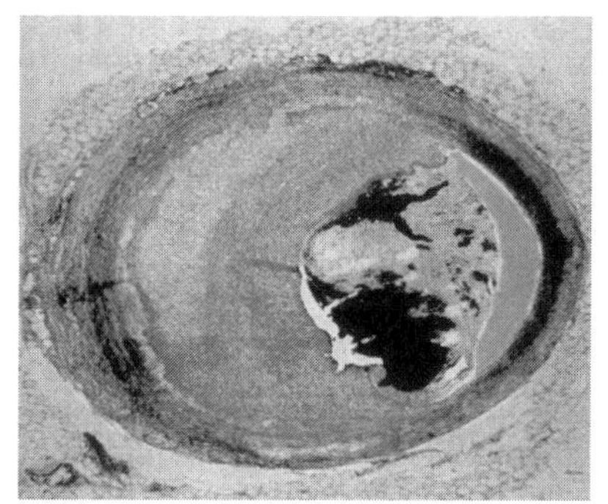

▲ **Figure 1.11b** Photomicrograph of a diseased coronary artery showing narrowing of the lumen due to build up of atherosclerotic plaque.

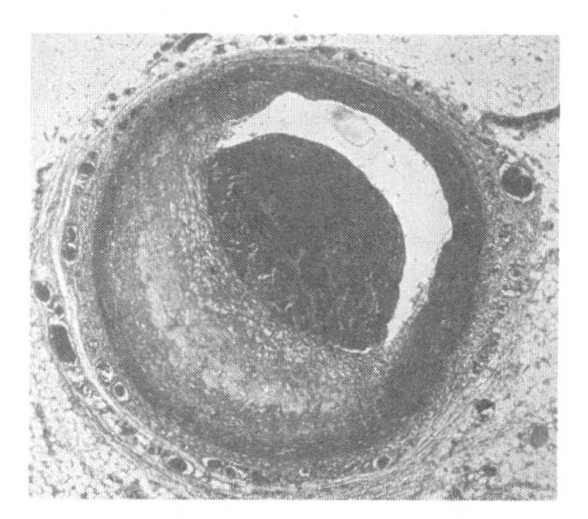

▲ **Figure 1.11c** Photomicrograph of a diseased coronary artery showing narrowing and a blood clot.

vessel wall and, in so doing, their cell surfaces change, causing them to stick to the wall and each other to form a platelet plug. They release fatty acids.

Damage of the vessel wall brings blood into direct contact with collagen within the wall. This triggers a complex series of chemical changes within the blood. A **cascade** of changes results in the soluble plasma protein called **prothrombin** being converted into **thrombin**. Thrombin is an enzyme that catalyses the conversion of **fibrinogen**, another soluble plasma protein, into long insoluble strands of the protein **fibrin**. These fibrin strands form a tangled mesh that traps blood cells to form a clot (Figure 1.12).

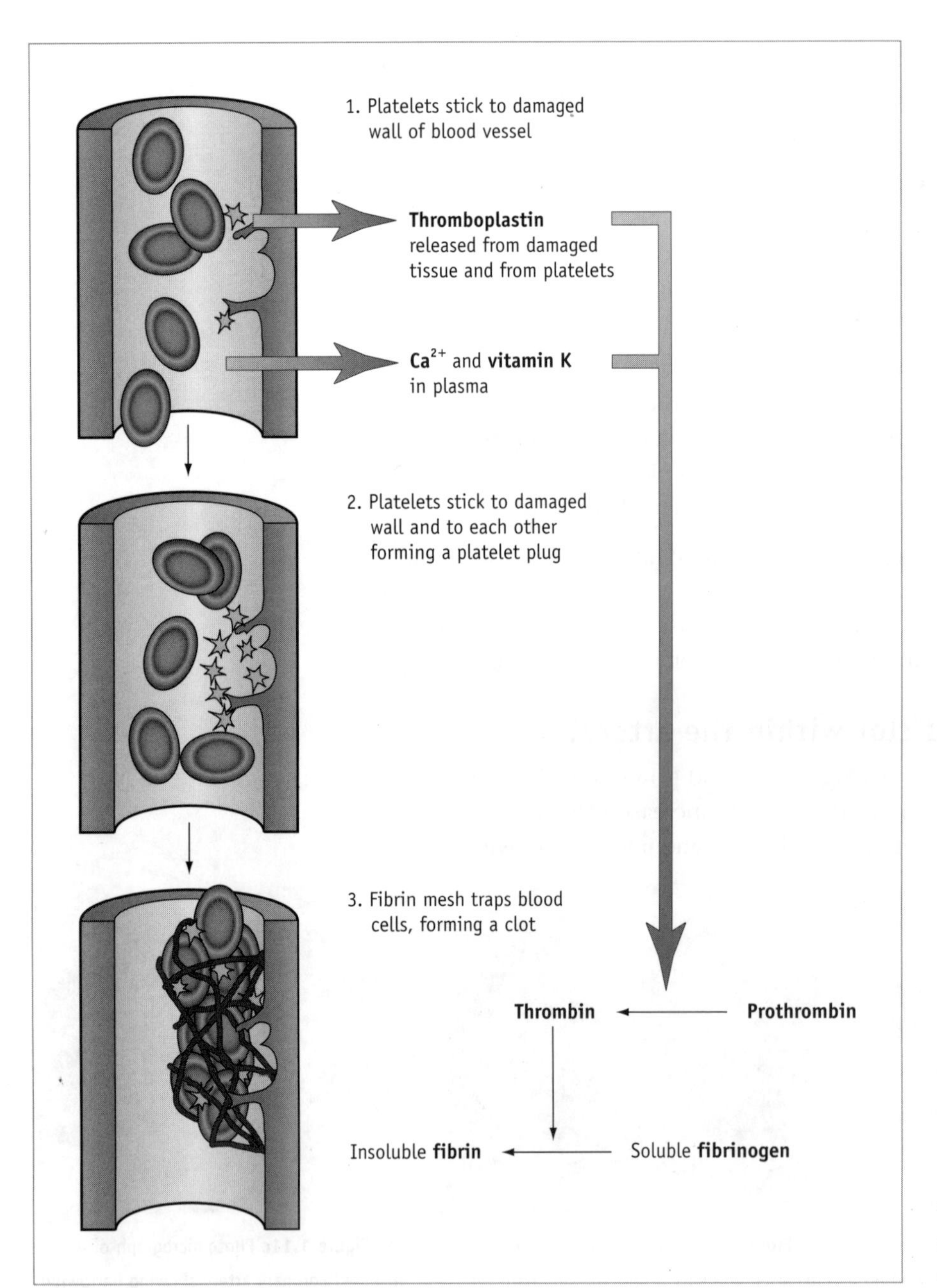

◀ **Figure 1.12** Damage to the vessel walls triggers a complicated series of reactions that leads to clotting.

Why do only arteries get atherosclerosis?

The fast flowing blood in arteries is under high pressure so there is a significant chance of damage to the walls. The low pressure in the veins means that there is little risk of damage to the walls.

A **haemorrhagic stroke** occurs when a blood vessel supplying blood to the brain bursts. If the burst occurs within the brain it is known as an *intracerebral haemorrhage*, whereas a vessel on the surface bursting causes what is known as a *subarachnoid haemorrhage*. Look at Figure 1.13 and work out why the different types of strokes were given these names. There is no need for you to remember these names but it is worth your being introduced to them as doctors and medical scientists use this sort of language.

Look back at Mark's story and decide which type of stroke he suffered.

 **Activity**

To check if you were right read Mark's full story in **Activity 1.6. AS01ACT06**

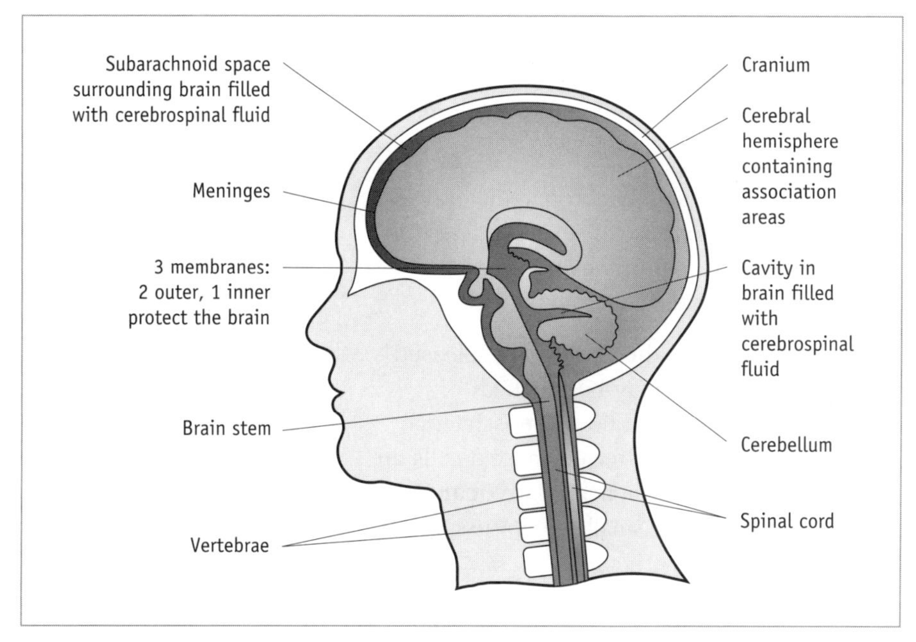

▲ **Figure 1.13** Blood vessels on the surface of or within the brain are susceptible to bursting. Bursting results in a stroke.

What are the symptoms of cardiovascular disease?

Stroke

The effects of a stroke will vary depending on the type of stroke, where the problem has occurred in the brain and the extent of the damage. The more extensive the damage the more severe the stroke and the lower the chance of full recovery. The symptoms normally appear very suddenly and include numbness, dizziness, confusion, slurred speech and blurred or loss of vision, often only in one eye. There is often paralysis on one side of the body with a drooping arm, leg or eyelid or a dribbling mouth as visible signs. The *right* side of the brain controls the *left* side of the body, and vice versa; therefore the paralysis occurs on the *opposite* side of the body to where the stroke occurred.

If the supply of blood to the brain is only briefly interrupted then a mini-stroke may occur, called a **transient ischaemic attack**. A transient ischaemic attack has all the symptoms of a full stroke but the effects only last for a short period and there can quite quickly be full recovery. However a transient ischaemic attack is a warning of problems with blood supply to the brain that could result in a full stroke in the future.

Coronary heart disease

When heart muscle does not receive enough oxygen-rich blood due to narrowing of the coronary arteries, the result may be a chest pain called **angina**. Angina is usually experienced during exertion, when the decreased blood flow to the heart muscle results in a lack of oxygen for the heart muscle, which is therefore forced to respire **anaerobically**. This produces **lactic acid** and the accumulation of lactic acid causes the pain of angina. Intense pain, an ache or feeling of constriction and discomfort is felt in the chest or in the left arm and shoulder.

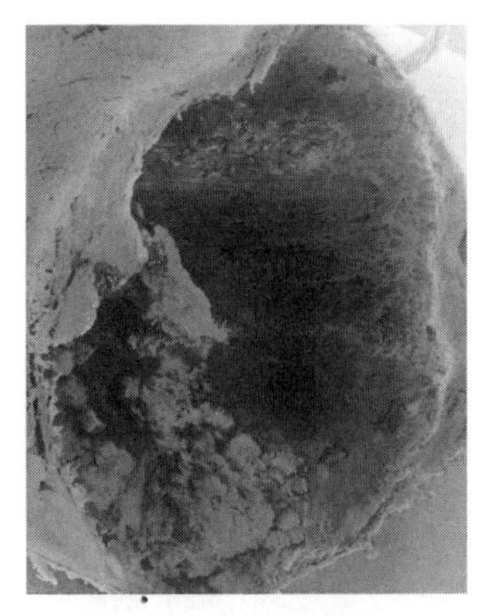

▲ **Figure 1.14** A small clot can block one of the coronary arteries.

Shortness of breath and angina are often the first sign of coronary heart disease. Other symptoms are unfortunately very similar to those of severe indigestion and include a feeling of heaviness, tightness, pain, burning and pressure – usually behind the breastbone, but sometimes in the jaw, arm or neck.

If a fatty plaque in the coronary arteries ruptures, the cholesterol released causes rapid clot formation and the blood supply may be blocked completely (Figure 1.14). Part of the heart muscle is said to be ischaemic (without blood). If the ischaemia is prolonged, the affected muscle cells are permanently damaged. This is what we call a heart attack or **myocardial infarction**. If the zone of dead cells only affects a small area of tissue the heart attack is less likely to prove fatal.

Sometimes the heart may also beat irregularly. This is known as **arrhythmia** and can itself lead to heart failure. It can be important in the diagnosis of coronary heart disease and you can read more details on page 20.

Narrowing of the arteries of the legs can result in gangrene and tissue death. More generally, blood can build up behind the portion of any artery that has narrowed and become less flexible; the artery bulges as it fills with blood and an **aneurysm** forms. An atherosclerotic aneurysm of the aorta is shown in Figure 1.15.

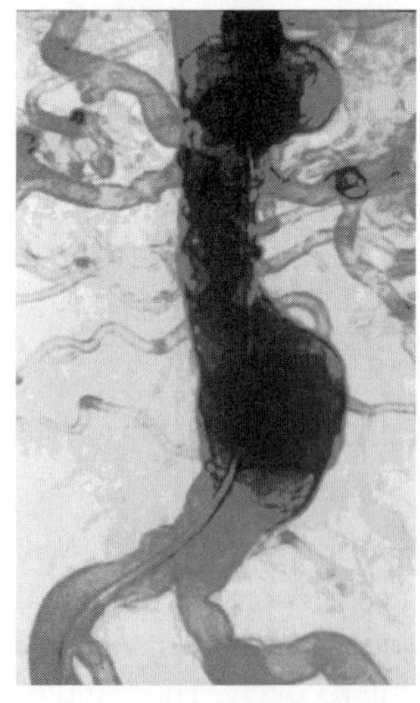

▲ **Figure 1.15** An aneurysm in the aorta like this one below the kidneys can be fatal if it ruptures.

What will eventually happen as the bulge enlarges and the walls of the aorta are stretched thin? Aortic aneurysms are prone to rupture when they reach about 6 to 7 cm in diameter. The resulting blood loss and shock can be fatal. Fortunately earlier signs of pain may prompt a visit to the doctor. The bulge can often be felt in a physical examination or seen with ultra-sound examination and it may be possible to surgically replace the damaged artery with a dacron graft (artificial artery).

If someone does have a heart attack, you may be able to save their life by carrying out cardiopulmonary resuscitation (often called 'artificial resuscitation').

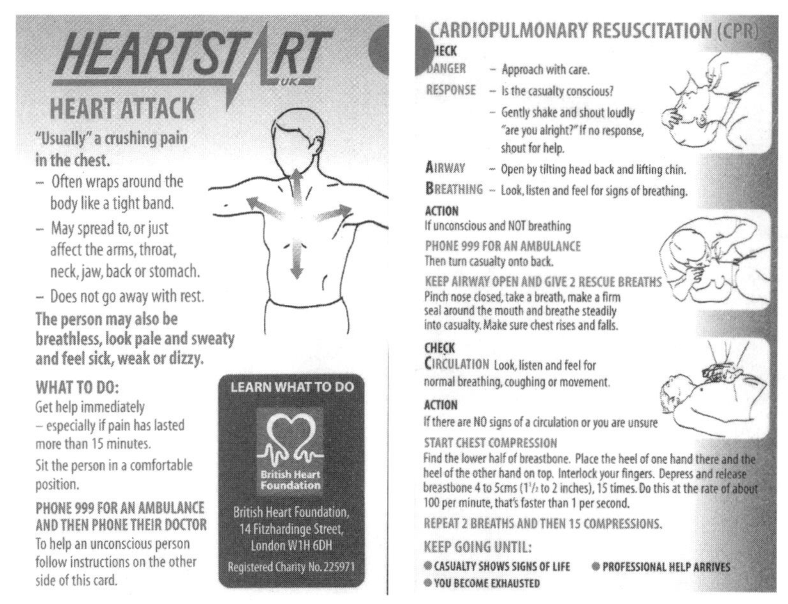

▲ **Figure 1.16** What would you do?

Activity

Complete **Activity 1.7** to find out more about cardiopulmonary resuscitation. **AS01ACT07**

How is cardiovascular disease diagnosed?

Following an examination of a patient suspected of having cardiovascular disease and a review of their medical history, a doctor will probably ask for tests to be carried out. The most commonly used test to check for problems with the heart is an **electrocardiogram (ECG)**. An electrocardiogram is also used on a patient with a suspected stroke to check for any heart condition that would increase the likelihood of stroke. If a stroke is suspected, as in Mark's case, a **computerised axial tomography (CAT) scan** and a **magnetic resonance imaging (MRI) scan** might be performed.

Extension

There are several other tests to diagnose cardiovascular disease that can be requested by doctors and you can read more details of these tests in extension 1 on the web site. **AS01EXT01**

CT or CAT scan

A computerised axial tomography scan, normally just referred to as a CT or CAT scan, uses a series of X-rays to build up an image of the brain. The computer generates a 3-D image from a series of X-rays taken as the scanner rotates around the patient.

Q1.8 Why is it better to rotate the scanner rather than the patient?

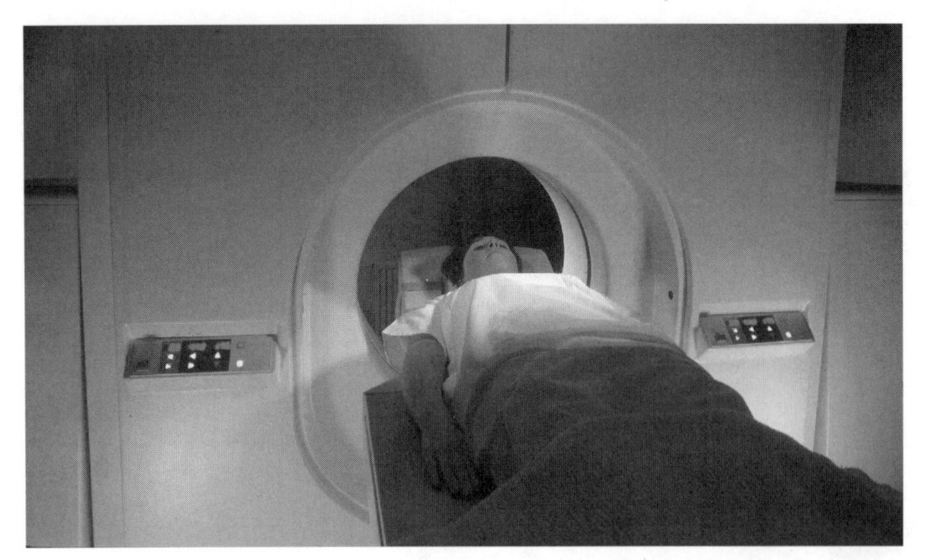

▲ **Figure 1.17** Mark was diagnosed using both CAT and MRI scans. This photograph shows a patient about to enter a CAT scan.

MRI

In a magnetic resonance imaging scan the patient is placed within a magnetic field and radio waves are used to create images of the brain or other parts of the body. The energy in the radio signals is absorbed by and then re-emitted by atoms within the body. The rate of emission is detected and is dependent on tissue type. The computer generates a series of black and white pictures representing different cross sections of the brain or other body part only a few millimetres apart. MRI images can therefore quickly reveal evidence of a stroke: areas of damaged brain tissue, aneurysms, internal bleeding or blockages.

MRI is also used for breast scans and the visualisation of injuries.

An electrocardiogram

An electrocardiogram is a graphic record of the electrical activity of the heart as it contracts and rests. It is easy to carry out and any patient with suspected cardiac problems will probably have one.

Contraction of heart muscle is initiated by small changes in the electrical charge of heart cells. When these cells have a slight positive charge on the outside, they are said to be **polarised**. When this charge is reversed, they are **depolarised**. A change in polarity spreads like a wave from cell to cell and causes the cells to contract. When there is a change in polarisation a small electrical current can be detected at the skin's surface. An ECG measures electrical current at the skin surface.

What is the normal electrical activity of the heart?

The heart will beat without any input from the nervous system and will continue to beat, even outside the body, as long as its cells are alive. How does it do this?

Depolarisation starts at the **sino-atrial node (SAN)**, a small area of specialised muscle fibres in the wall of the right atrium, located beneath the opening to the superior vena cava (Figure 1.18). The sino-atrial node is also known as the **pacemaker**. The SAN generates electrical impulses; the impulses spread across the right and left atrium causing them to contract at the same time. The impulse also travels to some specialised bundles of cells which take the impulse to the **atrio-ventricular node**, where conduction to the ventricles is delayed for about 0.13 s.

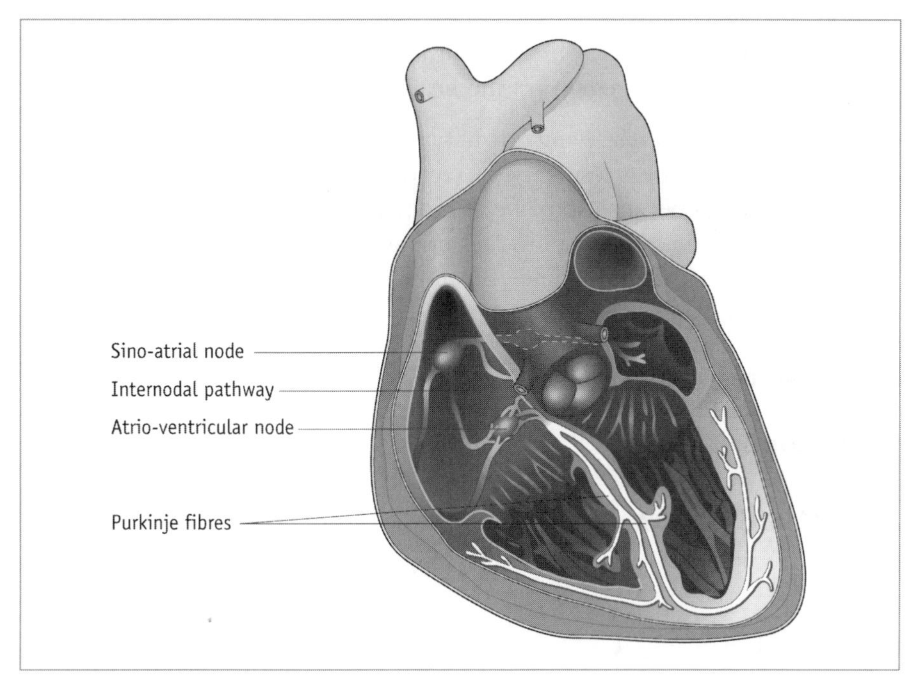

▲ **Figure 1.18** Follow the route taken by the electrical impulses passing over the heart and initiating contraction.

Why is it important that the impulse is delayed? The delay ensures that the atria have finished contracting and that the ventricles do not contract until they have filled with blood.

After this delay, the signal reaches the **Purkinje fibres**, large specialised muscle fibres that conduct impulses rapidly to the apex (tip) of the ventricles. There are right and left bundles of fibres and they are collectively called the **Bundle of His**.

The Purkinje fibres continue around each ventricle and divide into smaller branches that penetrate the ventricular muscle. They thus carry the impulse to the outside cells of the ventricles from where it spreads through each entire ventricle.

The first ventricular cells excited are at the apex of the heart, so that contraction begins at this point and travels *upwards* towards the atria. This produces the wringing contraction of the heart, pushing the blood into the aorta and the pulmonary artery.

Q1.9 Draw a flowchart to summarise the information that you have just read.

Activity

Activity 1.8 includes an animation of the heart's conductive pathways and lets you investigate its working in more detail. **AS01ACT08**

How is an ECG carried out?

In an ECG, leads are attached to the person's chest and limbs to record the electrical currents produced during the cardiac cycle, as the wave of depolarisation spreads across cells in different directions. To ensure a good reading, the skin is shaved and swabbed with alcohol and a conducting gel applied. The patient must remain still and relaxed throughout the procedure.

Twelve electrodes are used to give 12 views of the heart. Electrodes at right angles to the wave of depolarisation will detect little or no current. By using different combinations of electrodes, the ECG can detect electrical currents as they spread in different directions across different regions of the heart.

An ECG is usually performed while the patient is at rest, lying down, but it may be used in a **stress test** (a treadmill test). A stress test is used to record the heartbeat during exercise. It may be needed because some heart problems only emerge when the heart is working hard. The stress test involves doing an ECG before, during and after a period of exercise on a treadmill. Breathing rate and blood pressure may also be measured and recorded.

Q1.10 Why is the skin shaved and swabbed with alcohol?

Q1.11 What effect might exercise have on the ECG?

What does the ECG trace actually show us?

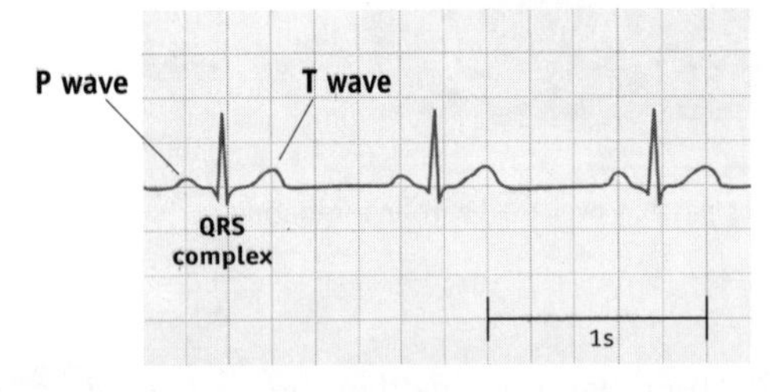

▲ **Figure 1.19** A normal ECG trace. The vertical axis indicates electrical activity; the horizontal axis indicates time.

Look at Figure 1.19. The P wave represents depolarisation of the atria and therefore their contraction. The QRS complex represents the wave of excitation in the ventricle walls. The PR interval is the time for impulses to be conducted from the atria to the ventricles, through the atrio-ventricular node. The T wave corresponds with the repolarisation (recovery) of the ventricles. This repolarisation is the heart's relaxation phase (diastole). An ECG does not show atrial repolarisation because the signals generated are small and are hidden by the QRS complex.

Using the ECG in diagnosis

During a period of ischaemia, the normal electrical activity and rhythm of the heart are disrupted, and arrhythmias (irregular beatings caused by electrical disturbances) can affect a larger area of heart muscle than that affected by the initial ischaemia (Figure 1.20).

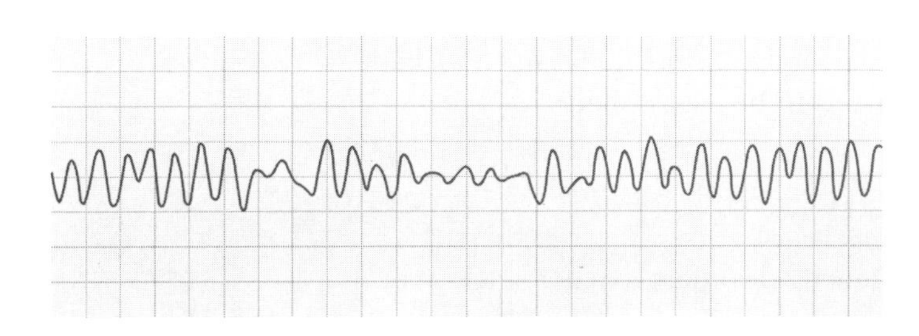

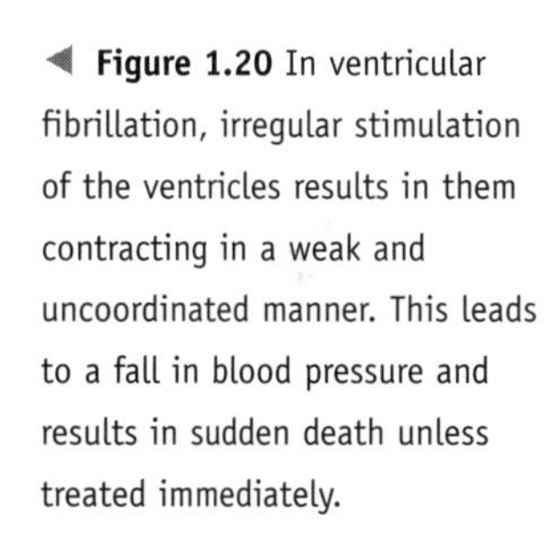

◀ **Figure 1.20** In ventricular fibrillation, irregular stimulation of the ventricles results in them contracting in a weak and uncoordinated manner. This leads to a fall in blood pressure and results in sudden death unless treated immediately.

Abnormal heartbeats, areas of damage and inadequate blood flow can all be seen in a patient's ECG. Look at Figure 1.21a which shows an ECG with an unnaturally long gap between the P wave and the QRS complex.

(a)

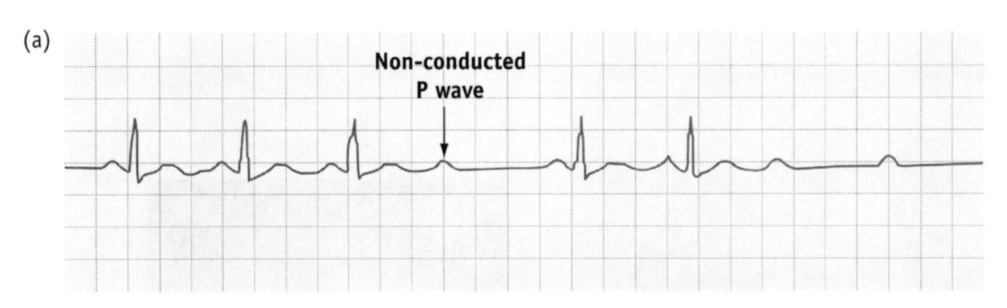

◀ **Figure 1.21a** Notice that the P wave is separated from the QRS complex, showing there is a break in the conduction system of the heart.

(b)

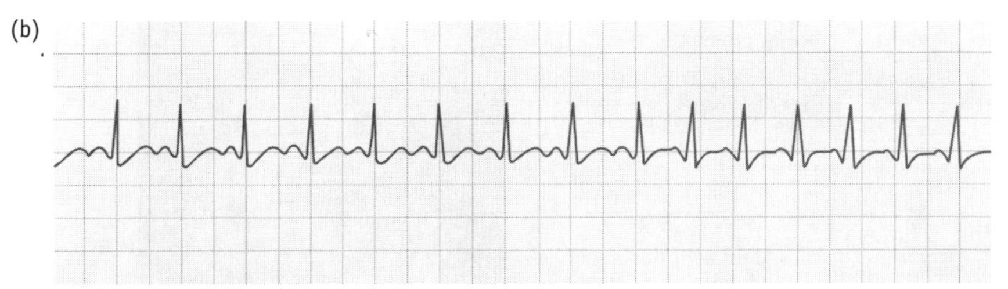

◀ **Figure 1.21b** This ECG trace shows abnormal rhythms that could lead to cardiac arrest.

Q1.12 In Figure 1.21a where do you think that the break in the conduction system occurs?

Q1.13 How does the trace in Figure 1.21b differ from that of a normal ECG?

Q1.14 The ECG in Figure 1.21b is from of a young woman who had collapsed at a club. Can you suggest what might have happened to produce these rapid rhythms?

Activity

Can you work out the patients' heart problem from the ECGs in **Activity 1.9**?
AS01ACT09

1.3 Who is at risk?

People are at risk from all sorts of things. In 2001, worldwide, there was one human heart attack every 4 seconds and a stroke every 5 seconds. In the UK the risk of having a fatal stroke in any one year is about 1 in 900 compared to 1 in 440 for a heart attack. However these probabilities use figures for the whole population giving averages which make the simplistic assumption that everyone has the same chance of having cardiovascular disease. This is obviously not the case.

The averages take no account of any **risk factors** – things that increase the chance of the harmful outcome. When assessing an individual's risk of bad health, all the contributing risk factors need to be established.

What are risk factors?

Some diseases, such as certain simple genetic disorders, have a single risk factor. Such diseases are 'determined' by inheritance of a single defective gene and the risk of inheriting the conditions follow the Mendelian rules of inheritance you have already met at GCSE.

For example, if two parents who are heterozygous for **albinism**, a genetic condition in which a defective gene prevents the production of pigment in skin or hair, have children what are the chances of their first child inheriting the condition? Figure 1.22 and the questions that follow should remind you abount the basics of inheritance.

Parent genotypes	Aa		Aa	
Parent phenotype	normal pigment		normal pigment	
Gamate	½ A	½ a	½ A	½ a

	A	a
A	AA	Aa
a	Aa	aa

any child with this genotype will be albino

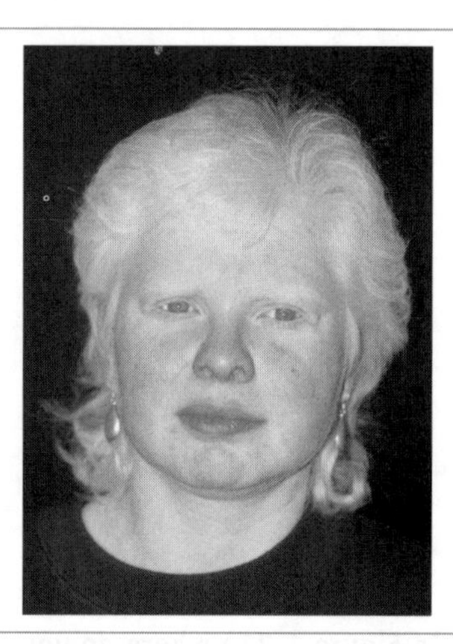

▲ **Figure 1.22** What are the chances of a child inheriting albinism from heterozygous parents?

Q1.15 Is the condition dominant or recessive? Give a reason for your answer.

Q1.16 What is the probability of a child inheriting the condition from parents who are:
a) both heterozygous
b) both homozygous recessive
c) both homozygous dominant?

In this situation and in some other genetically inherited conditions the risks are very clear cut. Often though the inheritance is much more complex. Even in some conditions that are controlled by a single gene, it is now known that different mutations of that gene determine the severity of the condition in the sufferer. This is true for cystic fibrosis that you will study in detail in topic 2.

Some diseases are the result of several genes interacting. In other diseases genes have been identified that do not *cause* the condition in the sense that there isn't a clear-cut relationship between having the genes and having the condition. Rather, the genes *increase* the individual's susceptibility to the disease.

The chances of a person suffering from one of the many common-health related problems, like cancer or cardiovascular disease, is rarely the consequence of genetic inheritance alone. Most of these diseases are **multifactoral**, with heredity, physical and social environment, and lifestyle behaviour choices *all* contributing to the risk. The combination of risk factors experienced by the individual exposes them to lesser or greater risk of the disease.

▲ **Figure 1.23** Professor Roy Anderson is an epidemiologist who has worked on such diseases as AIDS and foot-and-mouth disease.

How do scientists identify risks?

Epidemiologists, are scientists who study disease epidemics (Figure 1.23). One of the things they do is to compare groups of people who are suffering from a particular disease with groups without the disease. They collect a wide range of data; this might include age, weight, sex, lifestyle details, diet, exercise and family history. They then look for **correlations** between any of these factors and the occurrence of the disease. Large amounts of data are needed to ensure that the correlation is **statistically significant** in other words, not just an apparent correlation due to chance.

An example of such a study concerned the incidence of **toxic shock syndrome (TSS)**. 1980 saw a sharp increase in the number of cases of TSS, which is caused by a *Staphylococcus aureus* bacterium infection. Toxins released into the blood cause flu-like symptoms, which can be fatal. 38 deaths in the US prompted an epidemiological study to identify the risk factors. The US Center for Disease Control interviewed over 200 women and identified the use of tampons as a risk factor, with one particular make putting the user at an even greater risk. Removing this make from the market, combined with greater awareness among women of how to reduce the risk, saw a steep decline in the number of cases reported. A UK campaign in 1990 warned women of the risks and introduced a bill in the House of Commons to force tampon manufacturers to print warnings on tampon boxes. Warnings are now included in printed material within the packet.

An alternative approach to epidemiological study is to follow two groups of individuals over a long period of time. One group is selected from individuals exposed to a factor that is *suspected* of increasing the risk of contracting the disease under investigation, whereas the other group, the control group, is made up of individuals not exposed to this factor. Both groups are followed over a long period of time to determine if the members of the group exposed to the potential risk factor are indeed more likely to develop the disease. For example, a classic UK study of the incidence of lung cancer among groups of smokers and non-smokers identified smoking as the major risk factor for lung cancer.

Correlations and causations

It is important to appreciate that epidemiological studies don't *prove* that certain factors cause the disease in question. Consider the fact – and it is a fact beyond serious dispute – that people who smoke cigarettes for a number of years are more likely to develop lung cancer (not to mention heart disease and other diseases) than those who don't smoke. It *might*, logically, be the case that cigarette smoking *doesn't* cause lung cancer. Rather, it *might* be the case that, for example, there are some people predisposed (because of genetic factors, their upbringing or whatever) to start smoking cigarettes *and that these same people are more likely to develop lung cancer completely independently of whether or not they smoke cigarettes.*

This suggestion may sound far-fetched to you. But it illustrates a most important truth and that is that a correlation between two variables does not *necessarily* mean that the variables are *causally* linked. Indeed, it is easy to think of variables that are causally linked where there is no causation at all. For example, world-wide, speaking English as your first language probably correlates quite well with having a greater than average life expectancy. This, though, is simply because countries like the USA, UK, Australia and Canada have a higher than average standard of living and it is this that causes increased life expectancy through better nutrition, medical care and so on.

It is because of this logical gap between correlation and causation that scientists like, whenever they can, to carry out **experiments** in which they can **control variables**, to see if *altering* one variable really does have the predicted effect. As you will be well aware by now, scientists tend to set up a **null hypothesis**, assuming (for the sake or argument, as it were) that there will be *no* difference between an experimental group and a control group and then *testing* this hypothesis using statistical analysis.

Q1.17 How might a scientist, in theory, experimentally test whether cigarette smoking causes lung cancer in people?

Q1.18 Why would such an experiment be unethical?

Q1.19 Can you think of an experiment that might provide some experimental evidence as to whether or not cigarette smoking causes lung cancer?

Risk factors for cardiovascular disease

Your chances of having coronary heart disease or a stroke are increased by several inter-related risk factors, the majority of which are common to both conditions. Some of these you can control, while others you can't.

There have been a number of large-scale studies to identify risk factors for coronary heart disease, including the Seven Countries Study and the Framlingham Study. The World Health Organization MONICA (MONItoring, CArdiovascular disease) study, involving over 7 million people in 21 countries over 10 years, confirmed the earlier findings that linked several factors with increased incidence of the disease.

◀ **Figure 1.24** Some of the potential risk factors for developing coronary heart disease are easy to identify, but will altering behaviour to avoid risk factors reduce the chance of contracting the disease?

Does age and gender make a difference?

Q1.20 Look at Table 1.1. What happens to your risk of cardiovascular disease as you get older?

Q1.21 Does this mean that, at your age, you need not worry?

Q1.22 Do these data suggest that males and females face the same risk of cardiovascular disease?

Q1.23 Many people now think that, until menopause, a woman's reproductive hormones offer her protection from coronary heart disease. Do these data support this view? Is it valid to draw this conclusion from these data?

In **Activity 1.10** you use a spreadsheet to compare the data for coronary heart disease and stroke, look at trends over a ten year period and decide what was the risk of Mark having a stroke. **AS01ACT10**

Age (yrs)	Female deaths	Male deaths
0	48	68
1	17	13
2	4	6
3	6	2
4	1	1
5 to 9	12	4
10 to 14	10	19
15 to 19	24	34
20 to 24	42	74
25 to 29	102	137
30 to 34	149	255
35 to 39	222	579
40 to 44	422	1042
45 to 49	736	1978
50 to 54	1196	3677
55 to 59	2018	5223
60 to 64	3663	8554
65 to 69	7087	13382
70 to 74	13257	19618
75 to 79	22156	24504
80 to 84	27638	20881
85 +	56315	23067
Totals	135126	123114

▲ **Table 1.1** Mortality data from diseases of the ciculatory system for the UK in 1998

Source: World Health Organization (2001) http://www.who.ch/

My father had cardiovascular disease. Does that mean I will get it too?

If one or other of your parents suffer or suffered from heart disease, you are more likely to develop it yourself. There may be inherited predisposition for the disease.

Activity

You can read about the role of genes in the sudden death of athletes in **Activity 1.11. ASO1ACT11**

However, the inheritance of cardiovascular disease is not a simple case of a single faulty allele for the condition being passed from one generation to the next. There are several genes that can affect your likelihood of developing cardiovascular disease and these genes interact with each other and the environment to produce an overall effect. Some individuals may be in higher risk group because some of the factors run in families, for instance, high blood pressure and diabetes.

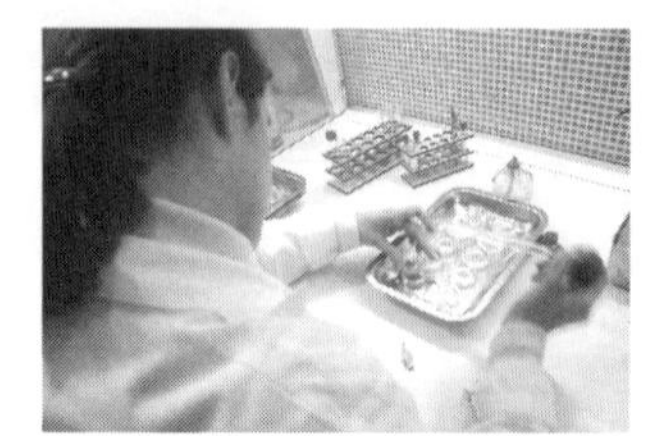

▲ **Figure 1.25** Genetic testing is becoming more common. Some companies now provide such testing so that individuals can see if they are at high or low risk of developing certain diseases.

Extension

The amount of cholesterol in your blood can also be affected by heredity. You can read more about inherited **hypercholesterolaemia** in extension 2. **ASO1EXT02**

There are now businesses that seek to meet people's concerns for the way in which their genetic endowment will affect their health and lifespan.

High blood pressure

Elevated blood pressure, known as **hypertension**, is considered to be one of the most significant factors in the development of cardiovascular disease.

Blood pressure is a measure of the pressure of the blood against the walls of a blood vessel. You should remember that blood pressure is higher in arteries than in veins, and is higher in veins than in capillaries. The pressure in an artery is highest during the ventricular contraction phase of the heart cycle, when the ventricles are contracting. This is the **systolic pressure**. Pressure is at its lowest in the artery when the ventricles are relaxed. This is the **diastolic pressure**.

Measuring blood pressure

A **sphygmomanometer** is a traditional device that measures blood pressure. It consists of an inflatable cuff that is wrapped around the upper arm and a gauge that measures pressure (Figure 1.26). Two values are obtained; the first number, the *systolic*, is the pressure obtained during systole, when the blood first manages to spurt through the artery that has been closed by the cuff. The second number, the *diastolic*, is the pressure of the blood flowing smoothly through the artery when the heart is not pumping. The SI units for pressure are kilopascals but in medical practice it is traditional to use millimetres of mercury, mmHg. (The numbers refer to the number of millimetres the pressure will raise a column of mercury.)

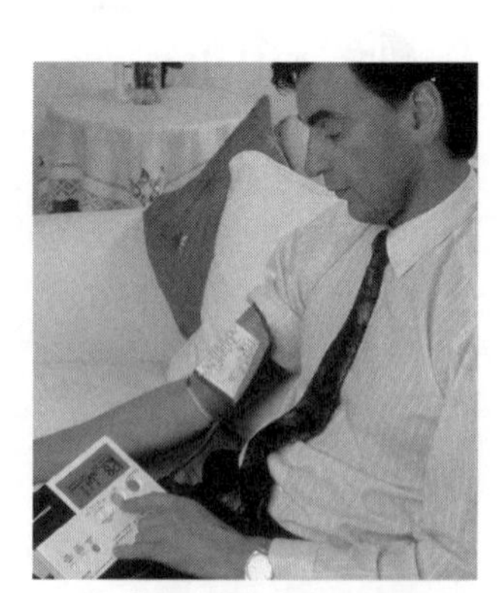

▲ **Figure 1.26** Nowadays blood pressure can be measured digitally by monitors that are easy to use.

Blood pressure is reported as two numbers, one 'over' the other, for example 120 / 80, meaning a systolic pressure of 120 mmHg and a diastolic pressure of 80 mmHg. For an average, healthy person you would expect a systolic pressure of between 100 and 140 mmHg and a diastolic pressure of between 60 and 90 mmHg.

Activity

In **Activity 1.12** you use a mercury sphygmomanometer or a digital blood pressure monitor to measure blood pressure. You could compare people's blood pressure in your class, or you could compare your own before and after exercise. **ASO1ACT12**

What determines your blood pressure?

Contact between blood and the walls of the blood vessels causes friction, and this impedes the flow of blood. This is called peripheral resistance. The greater total surface area in the arterioles and capillaries creates greater resistance to flow, slowing the blood down and causing the blood pressure to fall (Figure 1.27).

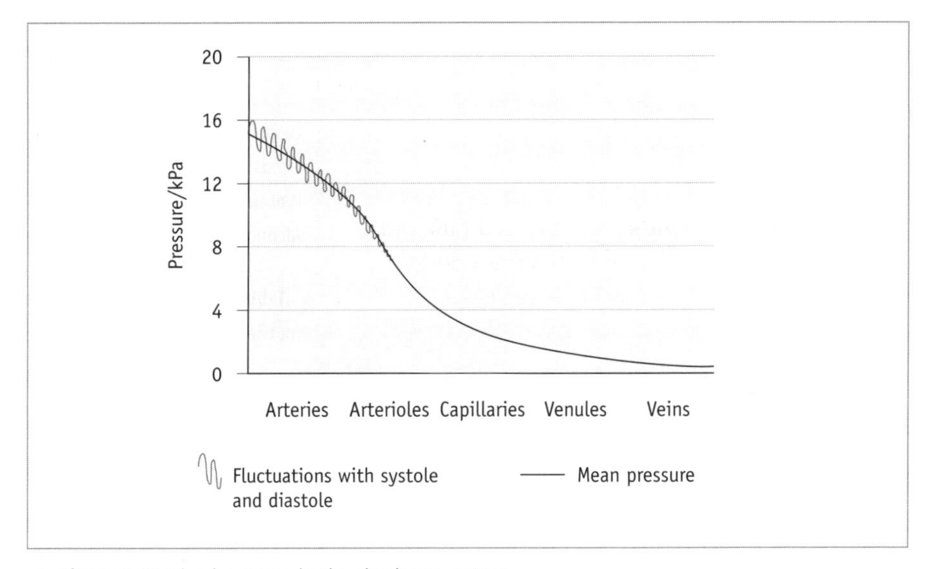

▲ **Figure 1.27** Blood pressure in the circulatory system.

If the smooth muscles in the walls of an artery or an arteriole contract they constrict, increasing resistance. In turn, your blood pressure is raised. If the smooth muscles relax, the lumen is dilated, so peripheral resistance is reduced and blood pressure falls.

Tissue fluid formation and oedema

At the arterial end of a capillary blood is under pressure which forces fluid out through the capillary walls into the intercellular spaces, forming **tissue fluid** (you will also see it called interstitial fluid in some books). Tissue fluid contains water and all the small molecules normally found in the plasma; the capillary walls prevent the passage of any blood cells and larger plasma proteins.

At the venous end of the capillary, due to the effect of peripheral resistance within the capillary, blood pressure is lower so fluid is no longer being forced outwards. Due to the loss of fluid, blood is more concentrated (there is less water per unit volume) and therefore fluid moves back into the capillary by **osmosis**. In addition, about 20% of the tissue fluid returns to the circulation via the **lymph vessels**. See Figure 1.28.

If blood pressure is elevated above normal, more fluid may be forced out of the capillaries. In such circumstances, fluid accumulates within the tissues causing swelling known as **oedema**. During left side heart failure (the most frequent type) there is an increase in pressure in the pulmonary vein and left atrium. This is because blood continues to flow out of the right side of the heart to the lungs and return to the heart due to the action of breathing muscles. There is a back up of blood in the pulmonary capillaries raising blood pressure. This causes accumulation of tissue fluid within the lungs impairing gas exchange.

Q1.24 What would happen in the case of right side heart failure?

Activity

Draw a **concept map** for blood pressure to bring together all the ideas covered. A pro-forma is available in **Activity 1.13** if you don't want to start from scratch. **AS01ACT13**

Dietary factors

Our choices of food, in particular the type and quantity of high energy-containing food, can either increase or decrease our risk of contracting certain diseases, including cardiovascular diseases.

Which nutrients store energy? **Carbohydrates**, **lipids** (often called fats and oils) and **proteins** are constituents of our food, all of which store energy. Alcohol can also provide energy. The relative energy content of these nutrients is shown in Table 1.2.

	Energy available per gram (kJ)
Carbohydrates	16
Lipids	37
Proteins	17
Alcohol	29

▲ **Table 1.2** Energy content of nutrients

Energy units – avoiding confusion

Most packet foods these days detail the energy content per 100 g or other appropriate quantity. In Figure 1.29 notice the units used to express energy content for the bag of crisps. Why are two different units used? Which should we prefer?

Traditionally, energy was measured in **calories**; one calorie is the quantity of heat energy required to raise the temperature of 1 ml of water by 1°C. Food labels normally display units of 1000 calories, called **kcalories** or **Calories**. The SI unit (International System of Units) for energy is the **joule** (J). 4.18 joules = 1 calorie. The **kilojoule** (1 kJ = 1000 joules) is used extensively in stating the energy contents of food. In the popular press the Calorie is used as the basic unit of energy, particularly with reference to weight control. Hence most food labels in the UK continue to quote both Calories and kilojoules.

Q1.25 A new-born baby requires around 2000 kJ per day. Express this (a) in calories; (b) in Calories.

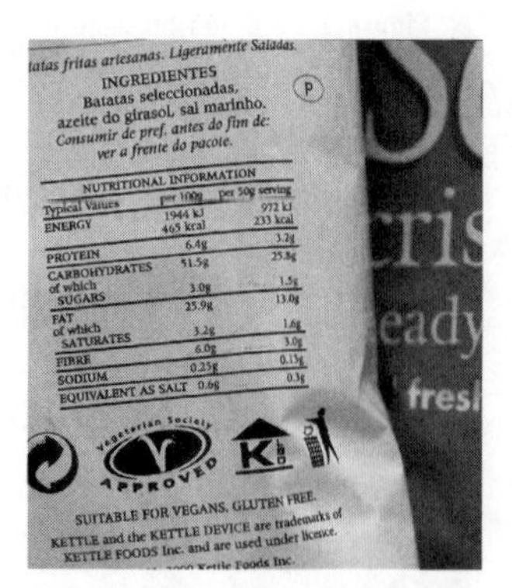

▲ **Figure 1.29** How much energy do these crisps contain?

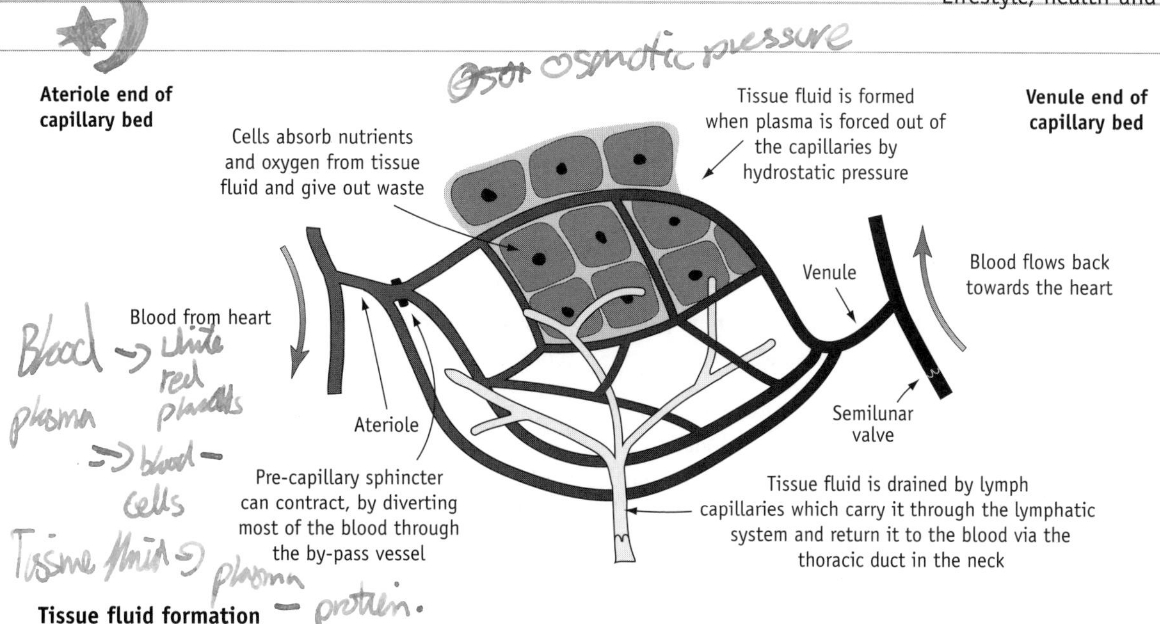

Tissue fluid formation

As plasma leaves the capillaries to form tissue fluid, dissolved substances such as oxygen and nutrients are carried out in the flow of fluid. This movement of fluid *out* of capillaries ay the arteriole end and then back *in* to capillaries at the venule end depends on two separate forces, one the result of hydrostatic pressure and one the result of osmotic pressure.

Hydrostatic pressure is the pressure exerted by a liquid, and this will be the same as 'blood pressure' in the vessels of the circulatory system. Blood pressure forces fluid out through the capillary walls. Tissue fluid bathing the cells also exerts hydrostatic pressure in the opposite direction, forcing water back into the capillaries. The *net* result of these opposing hydrostatic forces is shown as 'Net hydrostatic pressure' in the diagram.

Why is the hydrostatic pressure greater at the arteriole end of the capillary?

Osmotic pressure is the pressure which causes the movement of water from a less concentrated solution (i.e. more water) to a more concentrated solution (i.e. less water). Blood will always have a higher osmotic pressure than tissue fluid. This is due to the presence of plasma proteins, which are too large to flow out through the capillary walls with the other smaller, dissolved molecules. the blood and the tissue fluid exert osmotic pressure in opposite directions, but the *net* osmotic pressure will draw water into the capillaries from the tissues.

Why does the diagram show a greater net osmotic pressure at the venule end of the capillary?

The **net water movement** will depend on the balance between the hydrostatic and osmotic pressures of the blood and tissue fluid. Fluid leaves at the arteriole end of a capillary bed and re-enters the capillaries at the venule end. The net force causing tissue fluid formation is slightly greater than the force in the opposite direction at the venule end. In a normal situation the excess tissue fluid formed by this imbalance is drained away by the lymphatic system.

How would this diagram be different in the abnormal situation where excess tissue fluid formation results in oedema?

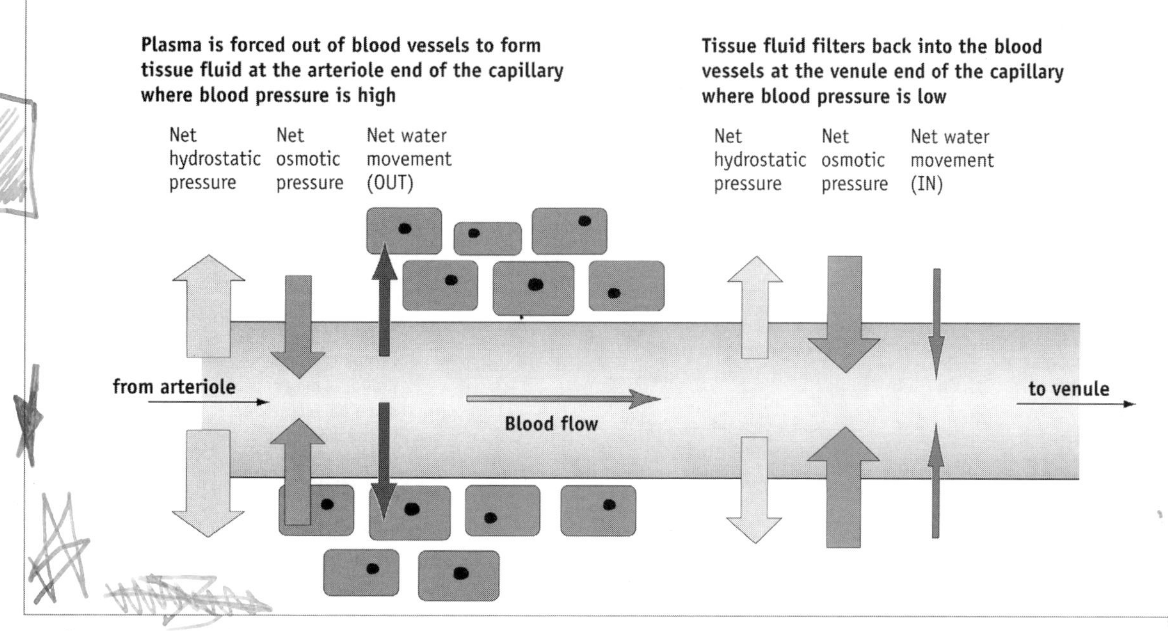

▲ **Figure 1.28** Production of tissue fluid in a capillary bed.

Nice to know: the role of culture in our eating habits

When a baby is born it survives entirely on milk (from the breast or a bottle) for a period that lasts from a couple of months to a year or more. Mother's milk is pretty much the same the world over. Aside from being a bit low in iron, fibre and some vitamins (e.g. vitamin C) it's a near perfect human diet.

Once **weaning** starts, a baby starts to take in solid food as well. This is where culture steps in. Adults across the globe eat very different foods and children, by and large, eat what they are given. The result is that we get used, as we grow up, to eating food that other people may think bizarre, even disgusting. Would you like to eat insect larvae, whale meat or sea cucumbers? Plenty of people do. Indeed, these foods are considered a delicacy in many countries.

There aren't just national differences in diet. In the UK, for example, there are regional differences – not just between England, Wales, Scotland and Northern Ireland but within each of these four areas too. Then there are differences related to social class (think about it!), age and gender. Marketing people know this only too well and carefully target food advertisements at the right 'segment' of the population.

And yet, one of the interesting things about what we eat nowadays is the result of **globalisation.** Nowadays most of us eat a greater variety of foods than our grandparents did. You probably eat Indian, Chinese and Italian foods, to mention just three nationalities. All in all we are remarkably adaptive as far as our diet goes.

Energy foods

Carbohydrates

The term carbohydrate was first used in the 19th century and means 'hydrated carbon'. If you look at each carbon in a carbohydrate molecule (Figure 1.30) you should be able to work out why.

Most people are familiar with sugar and starch being classified as carbohydrates but the term covers a large group of compounds with the general formula $C_x(H_20)_n$.

Activity

Complete the interactive tutorial in **Activity 1.14** to help you understand carbohydrate structure. **AS01ACT14**

Q1.26 What are the ratios of carbon, hydrogen and oxygen in carbohydrates?

Sugars are either **monosaccharides**, single sugar units, or **disaccharides**, where two single sugar units have combined in a condensation reaction, see Figure 1.30. Chains of between 3 and 10 sugar units are known as **oligosaccharides** and long straight or branched chains of sugar units form **polysaccharides**.

Glucose is important as the main substrate broken down in cellular respiration. Complex carbohydrates such as starch and glycogen are made up of glucose units joined together. When these are broken down in digestion, glucose is released, absorbed and transported to tissue cells in the bloodstream.

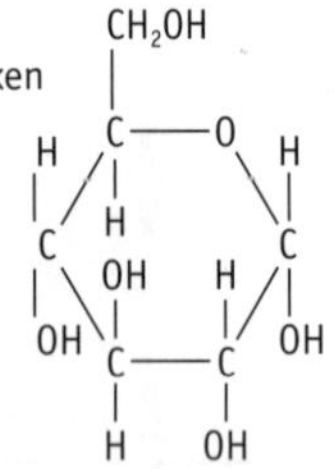

Glucose is a **hexose** sugar, because it contains 6 carbon atoms.

Fructose is the sugar which occurs naturally in fruit, honey and some vegetables. Its sweetness attracts animals to eat fruits and so help with seed dispersal.

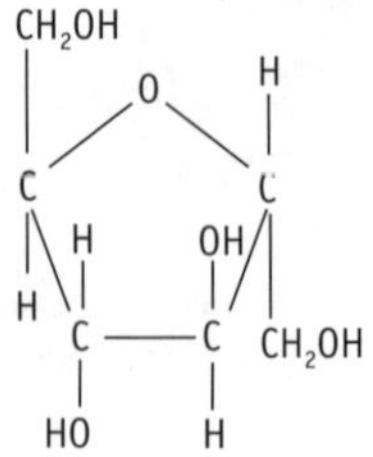

How many carbon atoms are there in fructose and galactose?

Galactose occurs in our diet mainly as part of the disaccharide sugar lactose. Fermented milk products may contain some galactose due to breakdown of the lactose during fermentation.

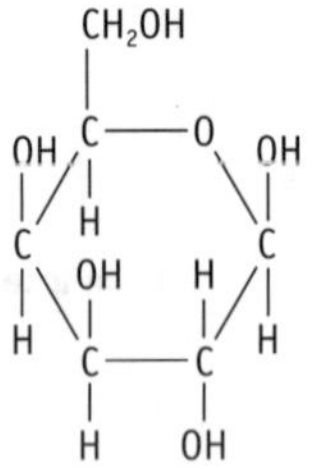

▲ **Figure 1.30** The structure of three nutritionally important monosaccharides.

Monosaccharides provide a rapid source of energy, being readily absorbed and requiring little, or in the case of glucose no, metabolism before use in respiration. **Glucose** and **fructose** are found naturally in fruit, vegetables and honey; they are both used extensively in cakes, biscuits and other prepared foods.

Common disaccharides found in food are **sucrose**, **maltose** and **lactose**. The sugar we use in cooking, in both white and brown crystalline form, and as golden syrup or molasses, is sucrose, extracted from sugar cane or sugar beet.

Activity

Activities 1.15 and **1.16** look at the hidden sugars in our food, and the perception of sweetness. **AS01ACT15, AS01ACT16**

Nice to know: why do we have such a sweet tooth?

We have taste receptors on the tongue for five main tastes – sweet, sour, bitter, salty and umami (the taste associated with mono-sodium glutamate (MSG)). It is likely that the sweet receptors enable animals to easily identify easily digestible food whereas bitter receptors provide a warning to avoid potential toxins. Humans, along with many other primates (apes and monkeys), have many more sweet receptors than most other animals. Our sweet receptors help us to identify when fruit is ready to eat.

Many adults are intolerant of lactose, the sugar present in milk. Drinking milk will produce unpleasant digestive problems for these people. One solution is to **hydrolyse** the lactose in milk, which converts the *disaccharide* lactose into the *monosaccharides* glucose and galactose.

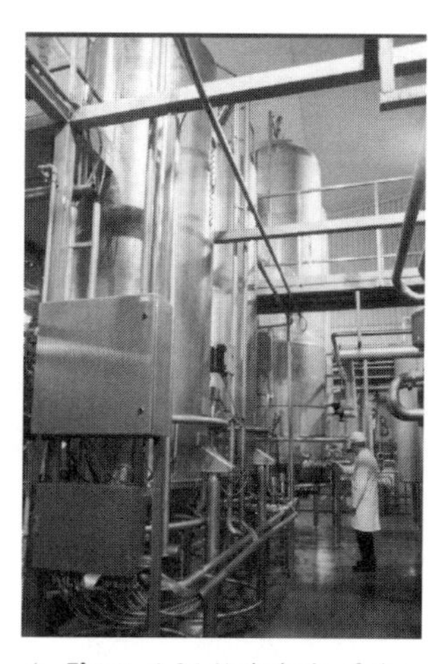

▲ **Figure 1.31** Hydrolysis of the lactose in whey waste, from cheese making, produces syrup used in the food industry.

Industrially this is carried out using the enzyme **lactase**. Lactase can be immobilised in a gel, and milk is poured in a continuous stream through a column containing beads of the immobilised enzyme (Figure 1.31). The specificity of enzymes (e.g. lactase) can make them a useful tool in the detection of certain chemicals.

As Asian and Afro-Caribbean people have a particularly high rate of lactose intolerance, the resulting lactose-free milk is then particularly suitable for food-aid programs. Untreated milk would cause further problems for people already suffering from malnutrition and dehydration.

Activity

In **Activity 1.17** you immobilise lactase and use it to hydrolyse lactose. **AS01ACT17**

Oligosaccharides are found in vegetables such as leeks, garlic, artichokes, lentils and beans. They are not easily digested and pass through the small intestine intact. Bacteria present in the large intestine then ferment them producing gases that cause flatulence.

There are three main types of polysaccharide found in food: **starch** and **cellulose** in plants, and **glycogen** in animals. Although all three are polymers of glucose molecules, they are sparingly soluble and do not taste sweet.

Two glucose molecules being joined in a **CONDENSATION** reaction to form the disaccharide **maltose.**

H_2O

Maltose

Disaccharides formed by joining two monosaccharide units:

Sucrose
Sucrose, formed from glucose and fructose, is the sugar used commonly by plants to transport from one region to another.

Lactose
Glucose and galactose make up **lactose**, the sugar found in milk. Some adults are lactose intolerant and have digestive problem if they drink milk.

Maltose is the disaccharide produced when amylase break down starch. It is found in germinating seeds such as barley as they break down their starch stores to use for food.

The **glycosidic link** between two glucose units can be split by **HYDROLYSIS**. In this reaction water is added to the bond.

H_2O

Hydrolysis takes place when carbohydrates are digested in the gut, and when carbohydrate stores in a cell are broken down to release sugars.

Disaccharides are broken down into monosaccharides, and starch and glycogen are broken down into shorter chains, oligosaccharides and eventually sugars.

Complex carbohydrates are digested and absorbed more slowly than simple sugars. For this reason they will not cause swings in blood sugar after a meal.

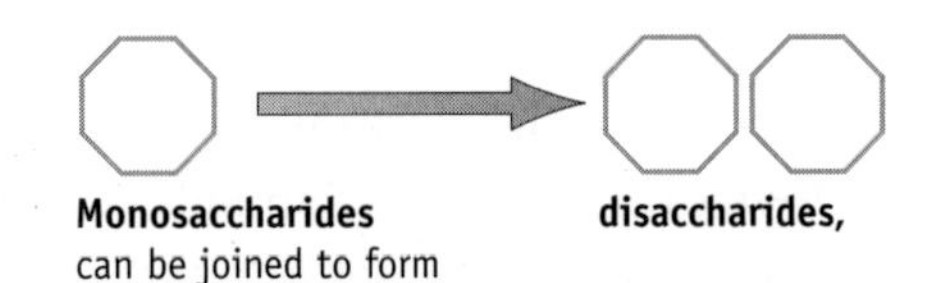

Monosaccharides can be joined to form **disaccharides,**

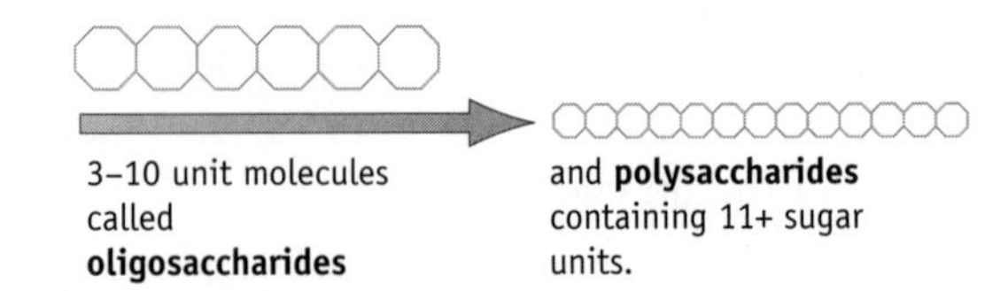

3–10 unit molecules called **oligosaccharides** and **polysaccharides** containing 11+ sugar units.

Oligosaccharides mainly escape digestion until the colon. Here, symbiotic bacteria break them down to shorter chains and sugars, often releasing hydrogen and carbon dioxide during this process. This explains why foods such as onions and beans can cause 'wind' during digestion.

▲ **Figure 1.32** Formation of polysaccharides.

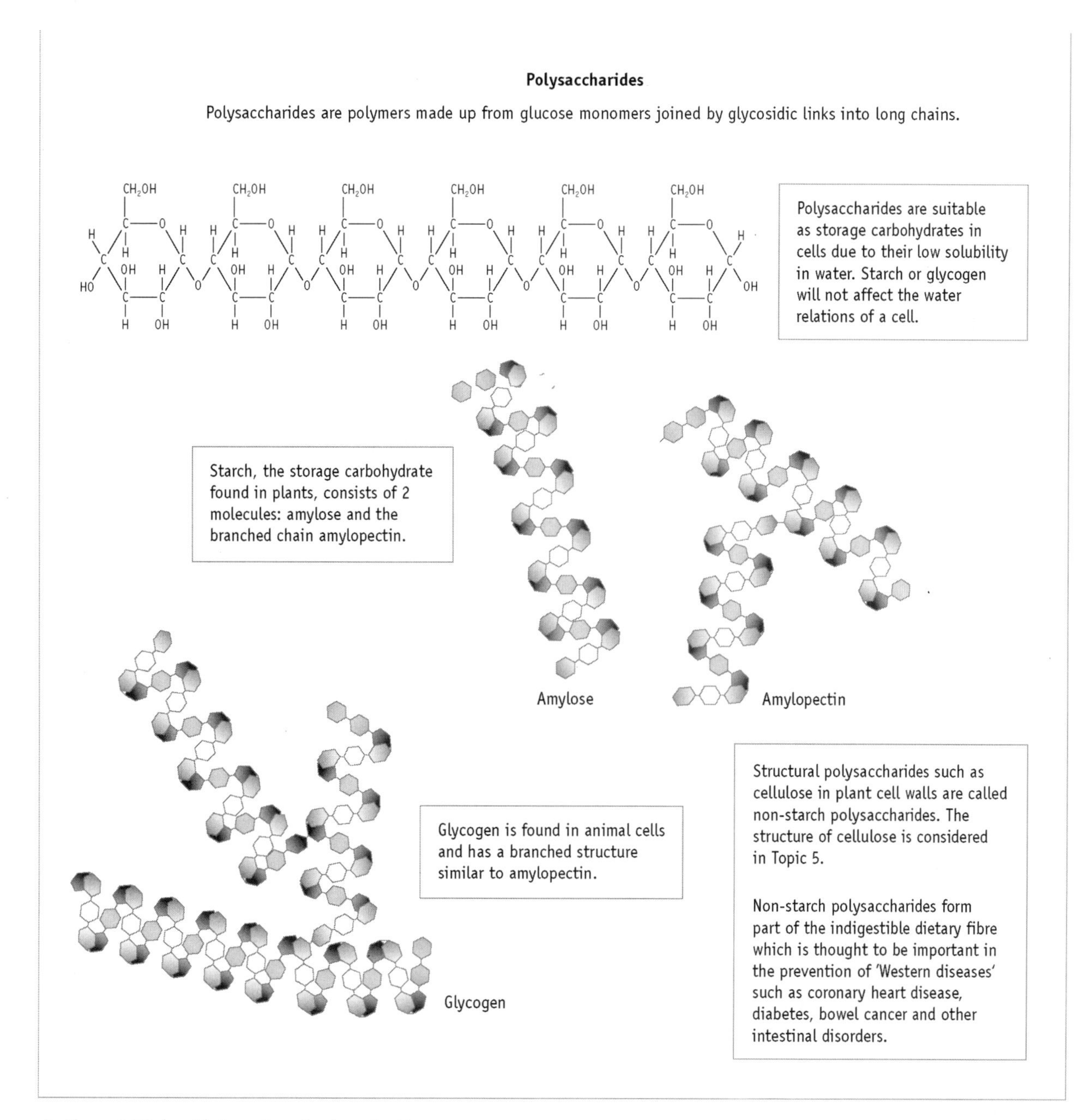

▲ **Figure 1.32 (cont.)** Formation of polysaccharides.

Starch is a major source of energy in our diet, and common in many foods (Figure 1.33). It occurs naturally in fruit, vegetables and cereals, often in large amounts. The sticky gel formed when it 'dissolves' in water makes it a good thickening agent and it is also added to many food products as a replacement for fat.

▲ **Figure 1.33** Foods high in starch.

Starch is made up of two molecules, amylose and amylopectin. **Amylose** is composed of a straight chain of between 200 and 5000 glucose molecules. The position of the bonds causes the chain to wind into a spiral shape. **Amylopectin** is also a polymer of glucose but it has side branches. Figure 1.32 attempts to show these complex 3D structures. Starch grains in most plant species are composed of about 70–80% amylopectin and 20–30%

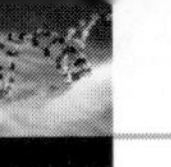

amylose. The compact spiral structure of starch and its insoluble nature make it an excellent storage molecule. It does not diffuse across cell membranes and has very little osmotic effect within the cell.

Bacteria and animals store glycogen instead of starch. Its numerous side branches (Figure 1.32) mean that it can be rapidly hydrolysed, giving easy access to energy stored in the liver and muscles.

Cellulose in the diet is known as **dietary fibre**, and it is also referred to as a non-starch polysaccharide. Up to 10 000 glucose molecules are joined to form a straight chain with no branches (the glucose molecules are a slightly different structure to those found in starch). Indigestible in the human gut, cellulose has an important function in the movement of material through the digestive tract.

Lipids

Lipids enhance the flavour and palatability of food making it feel smoother and creamier (Figure 1.34). They supply over twice the energy of carbohydrates, 37 kJ of energy per gram of food. This can be an advantage if large amount of energy need to be consumed in a small mass of food.

▲ **Figure 1.34** Which is more popular – with or without fat?

Lipids are organic molecules found in every type of cell; they are insoluble in water but soluble in organic solvents such as ethanol. The commonest lipids we eat are **triglycerides** (energy stores in plants and animals) – three fatty acids and glycerol linked by a condensation reaction (Figure 1.35). Other important lipids are **phospholipids** (components of cell membranes) and **steroids** (which include a number of hormones such as oestrogen and testosterone).

As well as supplying energy in the diet, fats also provide a source of **essential fatty acids**, those that the body cannot synthesise yet needs. In addition, fat soluble vitamins (A, D, E and K) can only enter our diet dissolved in fats. Fats must therefore be present in a balanced diet to avoid deficiency symptoms. For example, a deficiency of linoleic acid (an essential fatty acid) can result in scaly skin, hair loss and slow wound healing.

 **Activity**

Activity 1.18
Complete this ICT-based tutorial to understand lipid structure.
AS01ACT18

Triglycerides

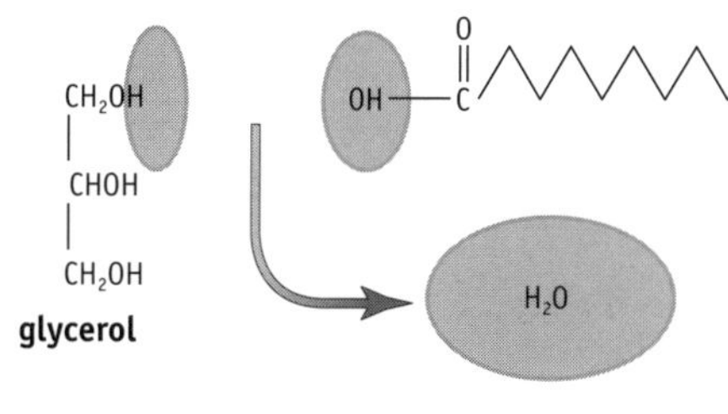

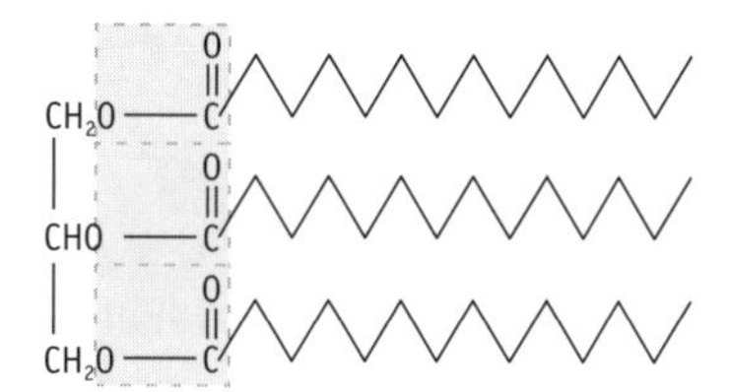

A **condensation** reaction removes water from between the glycerol and fatty acids to form **esther bonds**.

Fats and oils

The long hydrocarbon chain of a saturated fatty acid has all the carbon atoms saturated with hydrogens. Most animal fats have saturated fatty acid chains.

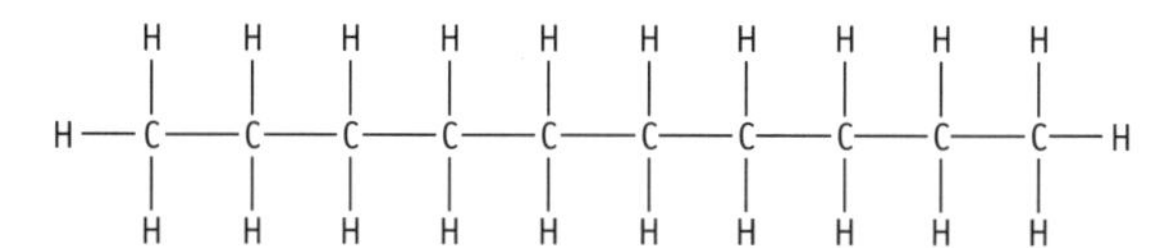

Monounsaturated fats such as olive oil have a single double bond between two carbon atoms.
Polyunsaturated fats like most other vegetable oils have a larger number of double bonds.

Palmitic acid is a saturated fatty acid with a straight hydrocarbon chain.

Oleic acid has a double bond which causes a kink in the hydrocarbon chain.

Straight, saturated hydrocarbon chains can pack together closely.
The strong intermolecular bonds between triglycerides made up of saturated fatty acids result in solid fats at room temperature.

Unsaturated hydrocarbon chains are unable to pack so closely together due to the kinks in the chains. The weaker intermolecular bonds between unsaturated triglycerides result in liquid oils at room temperature.

▲ **Figure 1.35** Structure and formation of lipids.

Key biological principle: The components of a balanced diet

A **balanced diet** contains:

- **water**
- **proteins** (including the **essential amino acids** we require and which the body cannot synthesise)
- **carbohydrates**
- **lipids** (including the essential fatty acids)
- **vitamins** (both water-soluble and fat-soluble)
- **minerals** (also known as **inorganic ions**) – e.g. Ca^{2+}, Fe^{2+} or Fe^{3+}, Cl^{-}
- indigestible **fibre.**

The energy balance

Look on food labels and you often find recommended daily amounts for nutrients, including daily energy requirements for men and women, but how much energy is right for each of us, and what happens if we don't get it right?

Getting it right

The UK Department of Health publishes dietary guidelines for most nutrients. They used to give **recommended daily amounts** but in 1991 these were largely replaced with **dietary reference values** (DRVs). These include: an estimated average requirement (EAR), a lower reference nutrient intake (LRNI) and a higher reference nutrient intake (HRNI) value, effectively providing a range of values within which a healthy balanced diet should fall. Upper and lower limits have not been set for carbohydrates and fats. Instead, estimated average requirements are suggested plus the average percentage of this energy that should come from the different energy component of a diet, table 1.3 gives the recommendations along with a recommendation for what percentage of the energy intake should come from eiter carbohydrates or fats.

Estimated average requirements for energy (EAR) for adults (kJ/day (Calories/day))			% of daily total food energy intake excluding alcohol		
Age (years)	Males	Females	Total fat (saturated)	Total carbohydrates (starch)	Non-starch polysaccharides
19–50	10 600 (2550)	8100 (1940)	35 (11)	50 (37)	15

▲ **Table 1.3** Dietary guidelines for carbohydrates and fats (HSMO 1991)

Activity

Activity 1.19 uses dietary analysis software for you to work out your energy budget and determine whether you are getting the right amount of energy and from the best sources. **AS01ACT19**

Getting it wrong

We have to be aware that we need both carbohydrates and fats in our diet for good health, but that there are consequences if we get it wrong by consuming too much energy or if the percentage supplied by the various components differs greatly from the guidelines.

The typical average energy requirement for men aged between 19 and 50 is 10 600 kJ and for women it is 8100 kJ. However, these are average values for the whole population and merely provide a guideline value to ensure that starvation and obesity are avoided. They do not give a precise recommendation for the amount of energy that is required by a particular individual.

You need a constant supply of energy to maintain your essential body processes, such as the pumping of the heart, breathing and the maintenance of a constant body temperature – processes that go on all the time, even when you are completely 'at rest'. The energy needed for these essential processes is called the **basal metabolic rate** (BMR) and varies between individuals. Males, for example, typically have higher rates than females. BMR (in units of energy needs per day) is higher with increasing body mass and falls with age; the decrease is greater for men than for women. The total energy you need also depends on how active you are. A person who cycles to work and enjoys playing squash will require more energy than someone who catches the bus and enjoys stamp collecting! The 'average' person may require between 8000 and 10 000 kJ a day but an athlete may require double this quantity. Some cyclists consume four times this amount.

If you routinely eat more energy than you use you have a positive energy balance. The additional energy will be stored and you will put on weight. In the UK, it is estimated that around 60% of men and 40% of women are either **overweight** or **obese**. Approximately 20% of the population are obese, with this figure trebling in the last 20 years. The increasing prevalence of obesity among both children and adults has been called an epidemic in Western Europe and the USA.

What do we really mean by 'overweight' and 'obese'?

Body mass index (BMI) is an internationally accepted method of classifying body weight relative to a person's height. To calculate BMI, body mass (in kg) is divided by height (in metres) squared, i.e.

$$\text{BMI} = \frac{\text{body mass (kg)}}{\text{height}^2\ (\text{m}^2)}$$

For example, the BMI of a person with a body mass of 65 kg and height of 1.72 m is 21.8. This figure can then be used to identify the category of body weight to which that person belongs, as shown in Table 1.4.

BMI	Classification of body weight
< 20	Underweight
20–24.9	Normal
25–29.9	Overweight
30–40	Obese
> 40	Severely obese

▲ **Table 1.4** The use of BMI to classify body weight

▲ **Figure 1.36** Over 5000 Calories a day and he still lost weight!

Q1.27 Calculate the body mass index of a person with body mass of 85 kg and height 1.68 m. How would you describe the body weight of this person?

A high fat diet will not necessarily result in weight gain if combined with a high level of physical activity. This was clearly illustrated by Ranulph Fiennes and Mike Stroud during their expedition to cross the Antarctica on foot – they found 5500 Calories a day inadequate to meet their energy needs (Figure 1.36).

A poor diet, particularly one high in fat, and a sedentary lifestyle are the major contributing factors to the development of obesity. There is evidence in the UK that fat consumption has actually declined since 1990 but greater inactivity means that obesity and associated conditions are on the increase.

Obesity increases your risk of coronary heart disease and stroke, even without other risk factors being present. The more excess fat you carry, especially around your middle, the greater the risk to your heart. Obesity raises your blood pressure and your cholesterol levels, and can greatly increase your risk for type 2-diabetes, too. Type 2-diabetes is also referred to as non-insulin dependent diabetes or late onset diabetes. It, in turn, increases your risk of coronary heart disease and stroke.

In type 2-diabetes the body either does not produce sufficient insulin or the body fails to respond to the insulin that is produced. You probably know that **insulin** is the hormone that helps regulate **blood glucose levels**. After a meal, the level of blood glucose rises; in response to this change the **pancreas** produces insulin, secreted into the bloodstream. It causes cells to absorb glucose, thus returning the level in the blood to normal. Continually high levels of blood glucose due to frequent consumption of sugar-rich foods can reduce the sensitivity of cells to insulin – type 2-diabetes results. It may take years to develop and may not even be diagnosed. It is thought that a million people in the UK are unaware that they have the type 2-diabetes.

Obesity also raises your blood pressure and elevates your blood lipid levels, two classic risk factors for cardiovascular disease. Epidemiological studies have shown a positive correlation between the percentage of saturated fat in the diet and high blood pressure. They have also shown a positive correlation between the percentage of saturated fat in the diet and increased incidence of cardiovascular disease. Much media attention, particularly in advertising, is focused on saturated fats and cholesterol.

What is the difference between saturated and unsaturated fat?

If fatty acid hydrocarbon chains contain the maximum number of hydrogen atoms they are said to be **saturated**; if there are any double bonds between carbon atoms, the fat is said to be **unsaturated** (figure 1.35). Saturated fats contain saturated fatty acids; **monounsaturated fats** contain fatty acids with a *single* double bond, whereas **polyunsaturated fats** contain fatty acids with more than one double bond.

Saturated fats are solid at room temperature. The fats from meat and dairy products are major sources of saturated fats. Olive oil is particularly high in monounsaturated fats. Most other vegetable oils, nuts and fish are good sources of polyunsaturated fats. Cholesterol is usually present alongside saturated fats.

Why are saturated fats and cholesterol such a problem?

Cholesterol (Figure 1.37) is an important and necessary component of cell membranes. The steroid sex hormones (e.g. progesterone and testosterone) and some growth hormones are produced from it. Bile salts, involved in lipid digestion and assimilation, are formed from cholesterol. For all these reasons, cholesterol is essential for good health.

Like all lipids, cholesterol is not soluble in water. In order to be transported in the bloodstream, insoluble cholesterol is combined with proteins to form soluble **lipoproteins**.

There is a considerable amount of epidemiological evidence to show that the higher your blood cholesterol, the greater your risk of coronary heart disease (Figure 1.38).

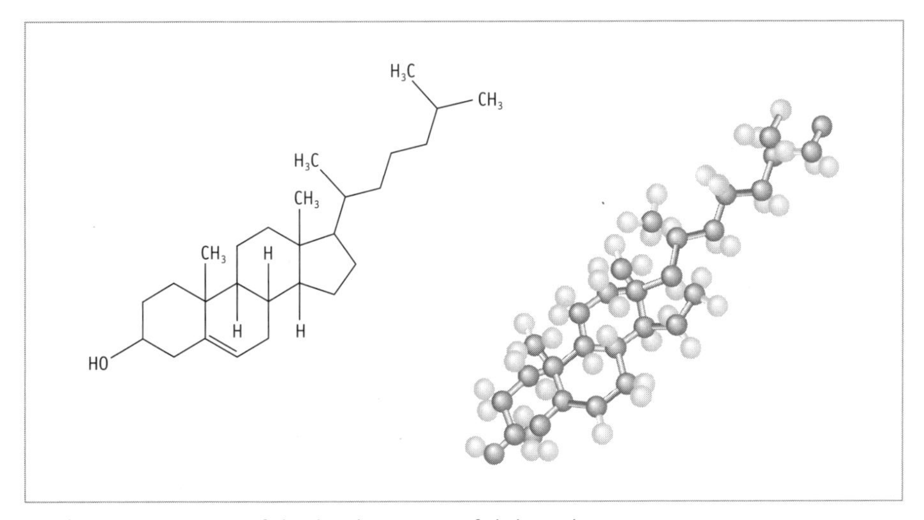

▲ **Figure 1.37** Two ways of showing the structure of cholesterol.

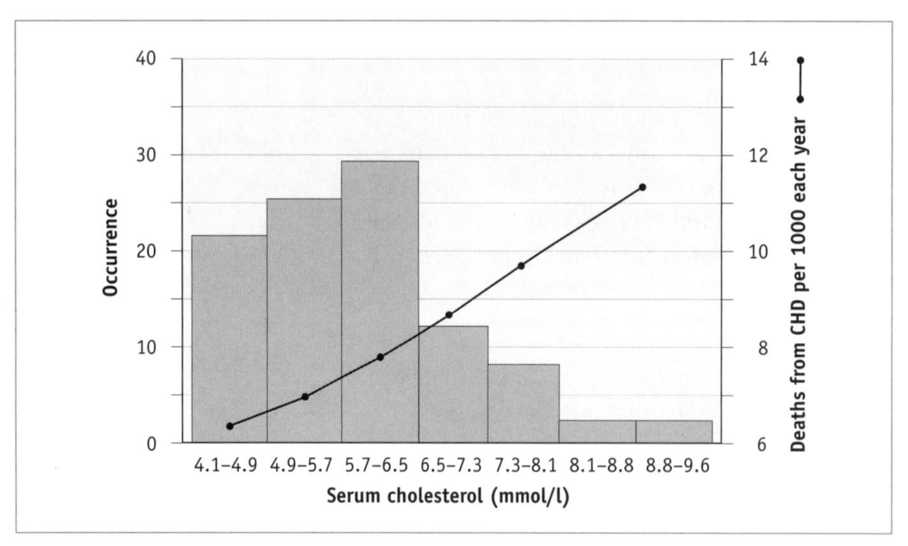

▲ **Figure 1.38** Blood cholesterol concentration related to coronary heart disease mortality in UK men aged 44–64 years. The joined up dots show the number of deaths from coronary heart disease with this serum cholesterol level per 1000 per year. The bars show the frequency of occurrence of the various serum cholesterol levels: for example, about 22% of UK men aged between 44 and 64 have a serum level of between 4.1 and 4.9 mmol per l.

Q1.28 Comment on the relationship between serum cholesterol levels and the risk of death from coronary heart disease.

It is estimated that in the UK 45% of deaths from coronary heart disease in men and 47% of deaths from coronary heart disease in women are due to a raised blood cholesterol level (> 5.2 mmol per l). It is thought that 10% of deaths from coronary heart disease in the UK could be avoided if everyone had a blood cholesterol level of less than 6.5 mmol per l.

However, as you may realise, it's not quite as simple as that! It is now thought that it may be not so much the total blood cholesterol level that is the risk factor, but the ratio between its two major transport lipoproteins – **high-density lipoproteins** (**HDLs**) to **low-density lipoproteins** (**LDLs**).

The high-density lipoproteins have a higher percentage of protein compared to LDLs, hence their higher density. The LDLs carry more cholesterol and less triglycerides than the HDLs.

The triglycerides from saturated fats in our diet combine with cholesterol and protein to form LDLs. LDLs circulating in the bloodstream bind to receptor sites on cell membranes before being taken up by the cells. Excess LDLs in the diet overload these membrane receptors, resulting in high blood cholesterol levels. Saturated fats may also reduce the activity of LDL receptors so the LDLs are not removed from the blood, thus further increasing the blood cholesterol levels.

High density lipoproteins are made when triglycerides from unsaturated fats combine with cholesterol and protein. HDLs transport cholesterol from the body tissues to the liver where it is broken down. This lowers blood cholesterol levels and helps remove the fatty plaques of artherosclerosis.

Monounsaturated fats are thought to help in the removal of LDLs from the blood. Polyunsaturated fats are thought to increase the activity of the LDL receptor sites so the LDLs are actively removed from the blood making the HDL to LDL ratio more favourable.

Mutations in genes that code for the LDL receptor protein cause familial hypercholesterolaemia. In people with familial hypercholesterolaemia tssue cells are unable to take up cholesterol. As a result, diet has to be carefully controlled to avoid high blood cholesterol levels and long-term medication may need to be taken.

LDLs are associated with the formation of atherosclerotic plaques whereas HDLs reduce blood cholesterol deposition. Therefore, it is desirable to maintain a high level of HDL (the so-called 'good cholesterol') and a low level of LDL (the so-called 'bad cholesterol').

Q1.29 Women generally have higher HDL-LDL ratios than men up until the menopause. What consequences would you expect this to have for the incidence of coronary heart disease in women related to men?

It is now possible to purchase a cholesterol testing kit (Figure 1.39) such as the Boots 3 minute whole blood cholesterol test. This test can help to identify elevated cholesterol levels associated with increased risk of coronary heart disease. However, these simple kits don't distinguish between LDLs and HDLs. It is possible for a laboratory to establish the LDL-HDL ratio by measuring the amount of cholesterol in each of the two fractions.

Activity

Activity 1.20 Are the foods that claim to be the 'low fat' or a 'healthy option' really what they seem? **AS01ACT20**

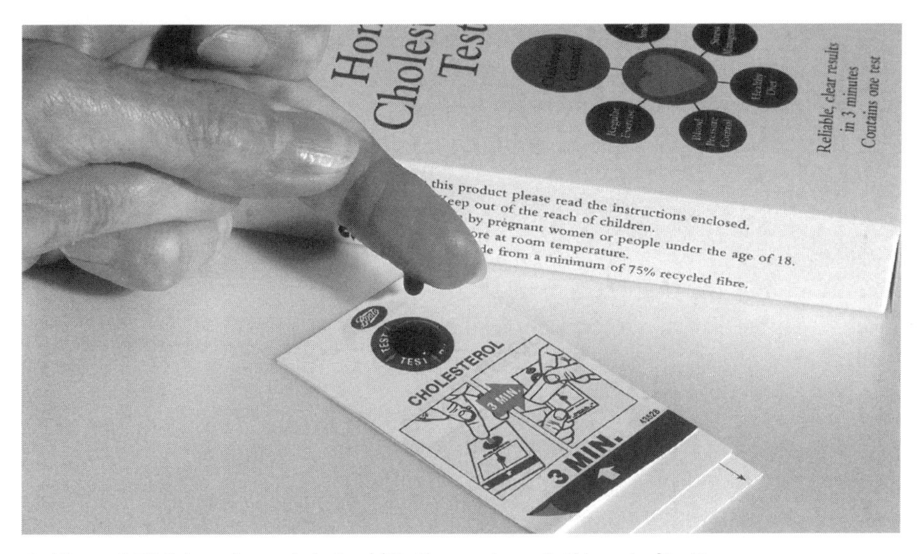

▲ **Figure 1.39** Using a home cholesterol kit. How much use is this sort of test?

Other factors that increase the risk of heart disease and strokes

The three most important factors that incease the risk of cardiovascular disease are smoking, having high blood pressure and having too high a level of LDLs (or LDL-HDL ratio). In addition, there are other things you should think about when deciding what to do if you want to lower your CVD risk.

Smoking, in addition to causing lung cancer, has the added disadvantage of introducing a vast range of toxins, including nicotine, into the blood which can damage artery walls and trigger atherosclerosis.

The role of antioxidants

During reactions in the body, unstable free radicals result when an atom has an unpaired electron, e.g. in the superoxide radical, $O_2^{\bullet}$. Free radicals are highly reactive and can damage many cell components including enzymes and genetic material. This type of cellular damage has been implicated in the development of some types of cancer, heart disease and premature ageing. Some vitamins, including vitamin C, beta-carotene and vitamin E, can protect against free radical damage. They provide hydrogen atoms that stabilise the free radical by pairing up with its unpaired electron. The MONICA study (mentioned on page 24) found that high levels of antioxidants seemed to protect against heart disease. Partly because of their function as antioxidants, the current recommendation is to include five portions of fruit or vegetables per day in our diet.

Activity 1.21 lets you determine if your diet contains enough antioxidant vitamins.
AS01ACT21

Q1.30 Why might it be better to obtain all the vitamins you need from basic foods rather than vitamin supplements?

Salt

The Food Standards Agency recommended salt intake for an adult is 6 g per day but the UK average intake is double that figure. 80% of the salt eaten coming from processed food; an average size bowl of cereal can contain as much as 1 g of salt. A high salt diet causes the kidneys to retain water. Higher fluid levels in the blood result in elevated blood pressure with the associated cardiovascular disease risks.

Inactivity

Exercise strengthens heart muscle so that it doesn't have to work so hard the rest of the time to pump blood around the body. Exercise also reduces the influence of other risk factors, such as diabetes, blood cholesterol and obesity. Sometimes it can actually lower blood pressure and it's certainly a great stress buster.

Studies show that the combination of exercise with weight loss is more effective than exercise alone for lowering blood pressure. Unfortunately, the majority of adults claim that they lack the time for exercise.

Stress

How you respond to stress in your life is very important. There is evidence that coronary heart disease is sometimes linked to poor stress management. Some people manage stress poorly, those with 'type A' personalities, for example (and some people cannot help this, for instance, those with a genetic predisposition to anxiety).

Nice to know: type A behaviour – how to get youself a heart attack

One day in the 1950s an upholsterer came to re-cover the waiting room seats at the office for two doctors in San Francisco. When he had finished he asked what sort of practice they had. The doctors said that they were both cardiologists. "I was just wondering", he replied, "because only the front edges of your seats are worn out".

One of the doctors was called Dr Meyer Friedman. The story is told in Friedman's prize-winning book *Type A Behaviour and Your Heart* which was published twenty years after the upholsterer's visit. It was the wife of a San Francisco executive who helped Friedman realise the link between stress and heart attacks. "If you really want to know what is giving our husbands heart attacks", she said, "I'll tell you. It's stress. The stress they get in their work. That's what's doing it." **Type A behaviour** is characterised by extreme competitiveness, by rushing to meet deadlines, by feeling harassed. Type As hate queues.

A lot of hard work by Friedman and others confirmed the upholsterer's insight. People with type A behaviour are more likely to suffer from a heart attack. But the reasons are complex and still not fully understood. One reason is that type As are, unsurprisingly, likely to have higher blood pressures – and high blood pressure, as we have seen (page 26) is associated with an increased risk of atherosclerosis and so eventually with a heart attack. Another reason is to do with changes in hormone levels in the blood. For example, type As, unsurprisingly, tend to show higher **adrenaline** levels, adrenaline being the hormone we produce when we are angry or frightened. One consequence of this is that arterioles constrict pushing blood pressure up (again); another is that triglycerides are mobilised into the bloodstream from fat stores (the body preparing us for 'flight or fight'). As a result of a complicated series of biochemical reactions these triglycerides end up carrying cholesterol to plaques in the walls of the arteries.

The net result is that suffering from too much stress isn't good for us. Relaxation, laughter and exercise are all good in themselves and effective ways of reducing stress and the chances of suffering a heart attack.

Do you think you are a type A personality or a more laidback type B who is less likely to get severely stressed?

Activity

In **Activity 1.22** is a teacher-led demonstration that lets you take part in an investigation of some factors that affect blood pressure and heart rate **AS01ACT22**

Q1.31 Do you think that some people do have a genetic predisposition to anxiety?

Q1.32 Discuss ways in which society could help such individuals.

Q1.33 Why does exercise reduce stress?

Drugs

The brain, composed of over 10 billion (10^{10}) nerve cells (**neurones**) each involved in up to 25 000 synapses, has a coordinating role. Specific areas within the brain control different functions. They receive **impulses** from **receptors** around the body, and through **synapses**, pass impulses to the **effectors** (muscles or glands) necessary for the desired response. Look at Figure 1.40. Follow the sequence of activities that transmit an impulse from one neurone to the next across the **synaptic cleft**.

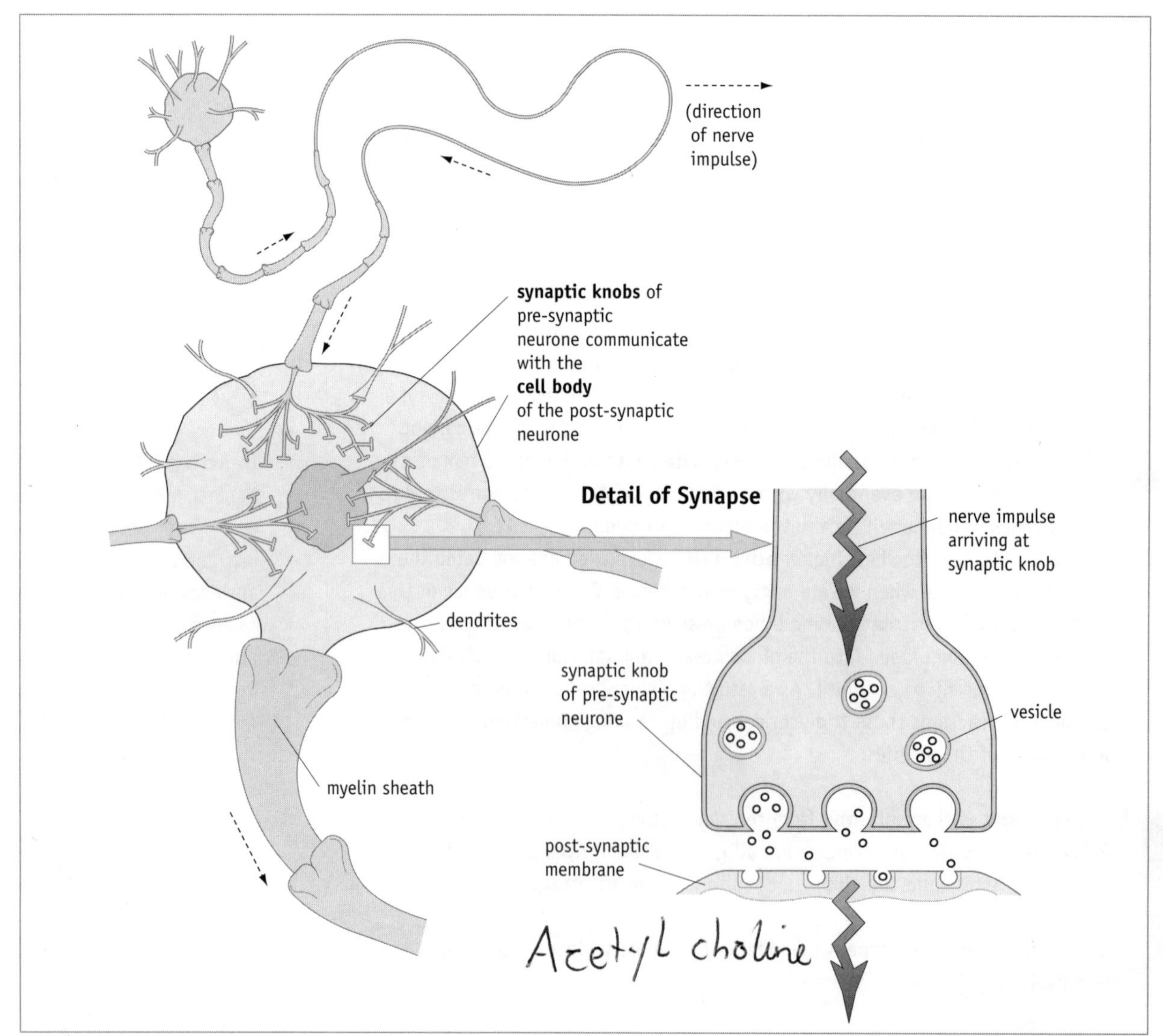

▲ **Figure 1.40** Chemical neurotransmitters transfer the impulse between adjacent neurones.

Acetyl choline ester

Psychoactive drugs affect synapses within the brain. Such drugs can be classified into two types: **excitatory (stimulant)** like caffeine, nicotine, cocaine and amphetamines (from which ecstasy is formed), and **inhibitory**, for instance alcohol, cannabis and opiates.

Alcohol

Heavy drinkers are at far greater risk of heart disease and a number of other diseases. Heavy drinking raises blood pressure, contributes to obesity and can cause irregular heartbeat. There has been much research and debate concerning potential *protective* effects of moderate drinking.

If you have a glass of wine the **alcohol** it contains, 1 unit or 8 g, is very quickly absorbed, 20% through the wall of the stomach and the remainder through the walls of the small intestine. (One unit of alcohol is approximately the amount that an adult eliminates from the body in one hour.) In the short term alcohol directly affects the brain. It acts on synapses. The alcohol binds to receptors on the **postsynaptic membrane** which makes **inhibitory neurotransmitters** bind more strongly to their respective receptors. These neurotransmitters prevent any impulse being generated in the postsynaptic neurone. This effectively closes these synapses and blocks these nerve pathways. Co-ordination between the sensory and motor areas of the brain is reduced.

Excess alcohol consumption can result in direct tissue damage, including damage to the liver, brain and heart. Such damage contributes to an increased risk of cardivascular disease. The liver has many functions but two of its main ones are processing carbohydrates, fats and proteins, and detoxification, including the removal and destruction of alcohol. High levels of alcohol can damage liver cells. This impairs the ability of the liver to remove glucose and lipids from the blood. In the liver alcohol is converted into acetaldehyde, a three carbon carbohydrate, and then metabolised. Most of the acetaldehyde is used in respiration but some ends up in very low density lipoproteins (LDLs). As they accumulate, these LDLs shift the HDL to LDL ratio, increasing the risk of plaque deposition.

Given these harmful consequences of alcohol, it seems remarkable to claim that moderate drinking may actually offer some degree of protection against cardiovascular disease. If this does turn out to be the case, it may not be the alcohol itself that protects but other constituents in the drinks. Wine and some fruit juices contain flavonoids and other constituent chemicals which have antioxidant properties and also reduce platelet aggregation.

Q1.34 How might the antioxidants in wine help reduce the incidence of cardiovascular disease?

If you do drink, moderation is the key! The UK recommended limits to avoid health problems are 2–3 units per day for women, and 3–4 units per day for men, with no binge drinking. There's one unit of alcohol in half a pint of average strength beer, a glass of table wine or a measure of spirits.

Q1.35 A former leader of the UK Conservative Party once claimed that he had sometimes drunk 15 pints of beer a night. How many units of alcohol would this be? What effect would you expect this to have on his health and capacity for sound judgement?

Activity

In **Activity 1.23.** you can find out if caffeine does increase heart rate and blood pressure. **AS01ACT23**

Activity

In **Activity 1.25** you can 'Test your healthy heart IQ'. **AS01ACT25**

1.4 Treatment for cardiovascular disease

If a patient is diagnosed with cardiovascular disease, the underlying problems that could trigger a stroke or heart attack need to be addressed. High blood pressure and high blood cholesterol are the conditions that might be treated. Lifestyle changes, such as stopping smoking, changing their diet and taking more exercise, would be recommended for many people.

Controlling high blood pressure

Many medications are available to lower high blood pressure. One class of drugs, called **diuretics**, increase the volume of urine produced by the kidney thus rid the body of excess fluids and salt. This leads to a decrease in plasma volume and cardiac output which lowers blood pressure.

Q1.36 We don't expect you to know much about how the kidneys work but can you suggest how diuretics might increase the volume of urine produced by the kidneys?

Another class of drugs called **beta-blockers** reduce the heart rate and the heart's output of blood. They do this by blocking nerve receptors for hormones, like adrenaline, that would normally directly affect the sino-atrial node. The effect of such hormones is to stimulate the frequency and strength of heart contractions with a resulting increase in heart rate and cardiac output. By blocking the receptors for these hormones, the beta-blockers reduce both heart rate and cardiac output.

Sympathetic nerve inhibitors are another important class of drugs known as **antihypertensives** (i.e. drugs that reduce high blood pressure). **Sympathetic nerves** go from the brain to all parts of the body, including the arteries. When activated, they can cause the arteries to constrict, raising blood pressure. This group of drugs reduces blood pressure by *inhibiting* these nerves and so preventing them from constricting blood vessels.

Anticoagulant and platelet inhibitory drugs

The tendency of blood to form **clots** is reduced by **platelet inhibitory drugs** and **anticoagulant drugs**. Aspirin is an anti-platelet drug; it reduces the 'stickiness' of platelets and likelihood of clot formation. A treatment involving a daily dose of aspirin and a second anti-clotting drug called clopidogrel has a dramatic combined effect. Trials involving 12 500 patients showed in 2001 that the combination reduces the risk of a patient dying from heart disease, having a repeat heart attack or suffering a stroke by 20 to 25 per cent. Another anticoagulant drug is **warfarin**. This can be taken orally for extended periods of time to prevent clotting.

Anticoagulant and platelet inhibitory drugs do not dissolve clots that have already formed. If a blockage has formed a clot-busting drug may be administered. **Streptokinase** is frequently use. Injected in to the patient the enzyme circulates within the blood and breaks down the clot.

Surgery

Those with severe coronary heart disease may need surgery. **Coronary angioplasty**, also known as **balloon angioplasty**, uses a catheter which is inserted into an artery in the groin and guided by X-ray imaging up to the narrowed coronary artery. A tiny balloon at the tip of the catheter is inflated and deflated to stretch or open the constriction and improve the passage for blood flow (Figure 1.41). The balloon-tipped catheter is then removed. The patient remains awake throughout the procedure.

In a **coronary artery bypass operation**, a blood vessel, usually taken from the leg or chest, is grafted onto the blocked artery, bypassing the blocked area. Two or three blocked arteries can be by-passed at once – a double or triple by-pass. The blood can then go around the obstruction to supply the heart with enough blood to relieve chest pain.

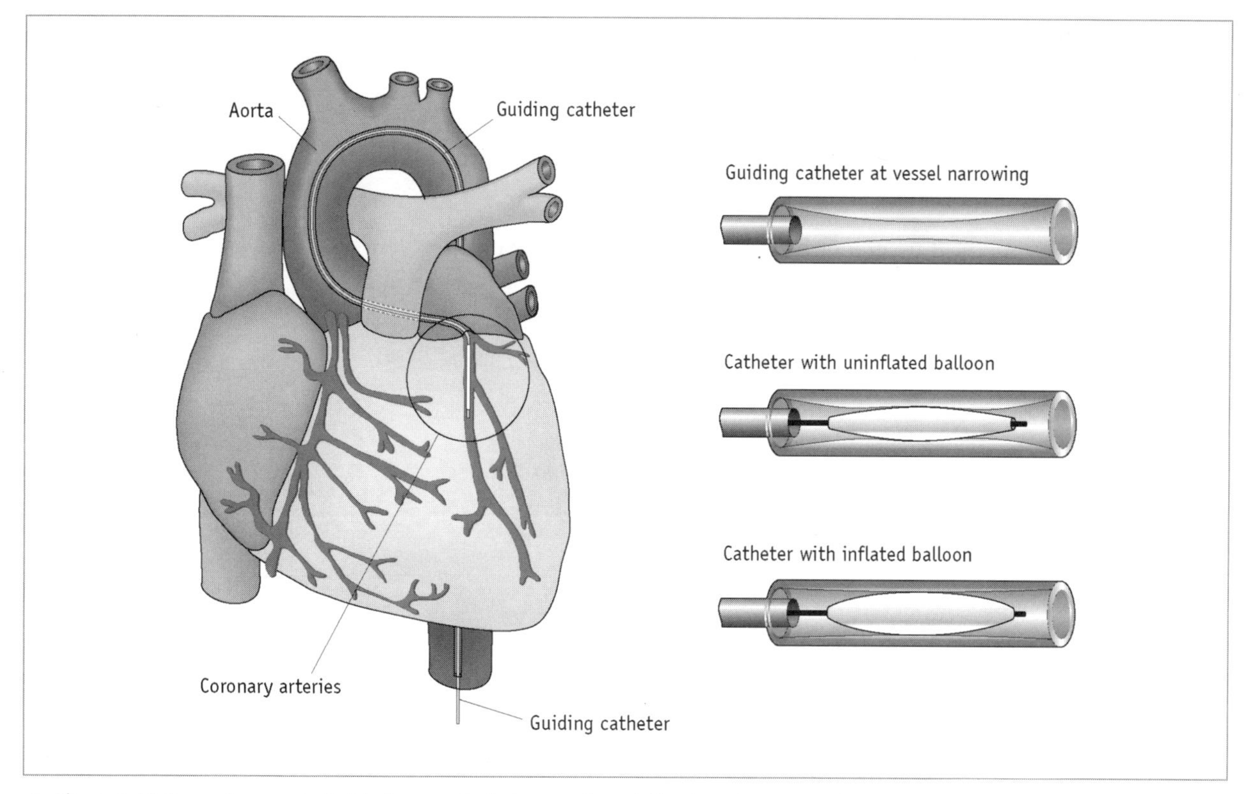

▲ **Figure 1.41** Research suggests that balloon angioplasty may give a better long-term survival rates than the use of drugs.

Other approaches

Research being undertaken into **gene therapy** suggests that injection of genes coding for a protein which enhances blood vessel growth may prove successful in helping to relieve some patients' symptoms.

Drugs may be administered to lower blood cholesterol and thus slow the development of artherosclerosis.

All in all, this is a truly exciting area of biomedical science and one in which new developments seem to occur daily. Try to keep abreast of these developments throughout your biology course, whether you are taking the one year AS course, or the two year AS and A2 course. You may find it helpful to read the extension material. 'New treatments for cardiovascular disease' provides a fine start. It even gives you the opportunity to observe surgery!

Extension

You can read more about new treatments for coronary heart disease in extension 3 on the website.
AS01EXT03

The way forward

The NHS National Service Framework for coronary heart disease has set out standards and services that should be available throughout England for people with diagnosed coronary heart disease. These include:

- advice about how to stop smoking (including **nicotine replacement therapy**)
- information about other modifiable risk factors
- advice and treatment to maintain blood pressure below 140/85 mmHg
- low dose aspirin (75 mg daily)
- dietary advice to lower serum cholesterol concentrations to either below 5 mmol per litre or by 30% (whichever is greater)
- beta-blockers for people who have had a myocardial infarction
- warfarin or aspirin for people over 60 years old who have atrial fibrillation
- meticulous control of blood pressure and blood glucose in people who have diabetes.

In addition to any medical treatment from cardiovascular disease, the patient will almost certainly need to make a number of changes in their lifestyle and these are considered in Section 1.5.

Activity

You can read some case studies in **Activity 1.26.** Decide for yourself what problem each patient was facing and recommend the appropriate treatment.
AS01ACT26

1.5 Reducing the risks of cardiovascular disease

If people in the UK did not smoke, the British Heart Foundation estimate that 10 000 fewer men and women of working age would die from heart attacks each year. Stop smoking and your risk of coronary heart disease is almost halved after only one year. Nicotine replacement therapy products are now available on NHS prescription.

26% of men and 13% of women consume more alcohol than the recommended 3–4 units per day for a man and 2–3 units for a woman. Research evidence suggests that drinking to excess, and in particular binge drinking, exposes individuals to the highest risks of cardiovascular disease.

There is evidence from Britain and America that untargeted general population cholesterol screening combined with dietary advice has little effect on lowering blood cholesterol levels. An Australian survey showed that 61% of people who had their blood cholesterol tested were unwilling to make changes on the grounds that their cholesterol was 'just right'. A further problem is associated with 'labelling' people. When told that they are hypertensive, many people react by signing off sick!

Q1.37 Why do you think some people react like when told their blood pressure is too high?

There is a need to reduce blood cholesterol in some people. One way to achieve this is through a low fat diet. The media constantly remind us of the need to do this.

The following data show the extent of cholesterol-lowering that is obtained from following a low fat diet in *high risk* patients, namely people who had already experienced a heart attack, compared with the general population.

	Blood cholesterol reduction	
	mmol per l	%
General population	0.22	3
High risk patients	0.65	9

▲ **Table 1.5** The effect of lipid lowering diets in reducing blood cholesterol levels

Q1.38 Can you suggest why the effect of the dietary intervention in patients who had experienced a heart attack appears to have been more successful in reducing blood cholesterol than in the general population?

Despite the greater percentage fall in blood cholesterol, surprisingly there was no significant fall in coronary heart disease *mortality* amongst the high risk group. One reason for this is thought to be that participants replaced lipids with complex carbohydrates in their diet.

Q1.39 How might replacing lipids with complex carbohydrates in the diet explain the fact that is no significant fall in coronary heart disease mortality amongst participants? (Hard!)

A diet to avoid cardiovascular disease

An energy balanced diet that avoids too much saturated fat and salt is recommended (Figure 1.42). The inclusion of some key dietary items may offer protection, as we shall now see.

▲ **Figure 1.42** A diet to reduce the risk of developing cardiovascular disease.

- **Soluble fibre**. This is known to lower blood cholesterol. Just two tablespoons of oat bran a day is effective!

Q1.40 Name some foods that are good sources of soluble fibre.

- **Oily fish**. Fish such as mackerel, sardines, anchovies, salmon and trout contain **omega–3 fatty acids** and these help to reduce clotting. The evidence for the importance of omega–3 fatty acids is seen in the Eskimos in Greenland and the inhabitants of certain Japanese islands. They regularly eat oily fish and have very low rates of coronary heart disease.

- **Vegetable oils**. Oils from corn, soy and safflower contain **omega–6 fatty acids** which lower LDL cholesterol.
- **Foods containing vitamin E**. Vitamin E can help to reduce blood cholesterol. It is an antioxidant and helps prevent damage to the blood vessel walls.

Q1.41 Name some foods that are good sources of vitamin E.

- **Fruit and vegetables**. Fruits and vegetables contain antioxidants and many also contain soluble fibre.

- **Garlic**. A compound in fresh garlic called allicin has been found in some studies to lower blood cholesterol. You can learn more from the activities in the extension material on Garlic.

- **Sterols and stanols**. These are naturally produced substances in plants. Structurally, they are very similar and resemble cholesterol. The difference between them is that **stanols** are produced by the hydrogenation of plant **sterols**. This means that stanols are *saturated*. Both sterols and stanols compete with cholesterol during its absorption in the intestine. Unfortunately, we would have to eat vast quantities of certain foods, such as vegetable oils and grains, in order to reduce our cholesterol

▲ **Figure 1.43** Pro-Active – a functional food.

levels through the activities of naturally occurring amounts of sterols and stanols. Industrialists have therefore worked hard to create products that can be incorporated at appropriate levels into everyday foodstuffs such as margarines. Controlled clinical trials of these **functional foods** show promising results: the amount of cholesterol absorbed from a food such as Pro.Activ is about 20% compared with 50% absorption from a food without stanols.

Activity

The Pro.Activ story in **Activity 1.27** lets you can find out more about these trials. **AS01ACT27**

Summary

Having completed Topic 1 you should be able to:

- Explain why many animals have a heart and circulation.
- Explain how the structures of blood capillaries, arteries and veins relate to their functions.
- Describe the symptoms of cardiovascular disease (CVD) i.e. coronary heart disease (CHD) and stroke.
- Describe what is meant by blood pressure and explain the significance of high blood pressure in cardiovascular disease.
- Explain how oedema arises from abnormal tissue fluid accumulation.
- Relate the structure and operation of the mammalian heart to its function.
- Describe the normal electrical activity of the heart and how this can aid the diagnosis of CVD and other heart conditions through electrocardiograms (ECG).
- Explain the course of events that leads to atherosclerosis (endothelial damage, inflammatory response, plaque formation, raised blood pressure).
- Describe the blood clotting process (thromboplastin release, conversion of prothrombin to thrombin and fibrinogen to fibrin) and its role in cardiovascular disease.

- Describe the factors which increase the risk of CVD (genetic, diet, age, gender, high blood pressure).
- Explain the types of experimental methods (control of variables and null hypotheses) undertaken by epidemiologists in their studies to determine the risk factors associated with human health.
- Analyse quantitative data, on illness and mortality rates, to determine health risks and recognise that it is important to distinguish between correlation and causation.
- Explain why people's perceptions of risks are often different from the actual risks.
- Describe how enzymes can be immobilised for use and explain the advantages of immobilising enzymes.
- Analyse data on energy budgets and diet.
- Distinguish between monosaccharides, disaccharides and polysaccharides (glycogen and starch) in terms of their structure and their role in the diet.
- Describe how monosaccharides can join to form polysaccharides through condensation reactions.
- Recognise the structures of and differences between saturated and unsaturated lipids.
- Calculate body mass indices (BMIs) and explain their significance.
- Discuss the possible significance for health of blood cholesterol levels and high-density lipoprotein to low-density lipoprotein ratios (HDL:LDL ratios).
- Describe the roles of drugs in the treatment of CHD (inhibition of blood clotting, inhibition of nervous stimulation of the heart and the use of diuretics to reduce tissue fluid retention).
- Describe how the effect of caffeine on heart rate in daphnia can be investigated practically.
- Discuss how individuals, by changing their diet, taking exercise and not smoking, can reduce their risk of CHD and summarise the importance of antioxidants in human diets.

Review test

Now that you have finished Topic 1, complete the end-of-topic test before starting Topic 2.
AS01RVT02

Genes and health

Genes and health

Topic 2

Why a topic called *Genes and health*?

It is now recognised that our genes have a major part to play in our health. Nearly 10 000 genetic disorders have been described. Some are very minor and create little or no problem; for example, colour blindness and hairy ears are both inherited conditions. Others have more serious consequences for the affected individual. Albinism, cystic fibrosis, Huntington's disease, sickle cell disease and haemophilia are all serious conditions caused by faulty genes. Most serious genetic disorders are rare, thankfully. But when one does occur it has a significant effect on the person's life and on the lives of their family. Not only must affected individuals deal with the condition itself, they may also face difficult decisions about the possible inheritance of the condition by the next generation.

Cystic fibrosis dilemma?

Most couples planning to start a family will have to take many decisions. Should it be a home or hospital birth, with or without pain relief? What should we call him or her? What colour should we paint their bedroom?

In this topic we follow Claire and Nathan who face a more daunting dilemma – should they have a child that could have cystic fibrosis (Figure 2.1)? Cystic fibrosis (CF) is one of the most common genetic diseases. Claire's Mum has the condition so they know it might be passed on to their child. To make the decision they need more information. If the child does inherit cystic fibrosis how will it be affected? How is cystic fibrosis inherited? What are the chances of any child they have inheriting the condition? Can genetic screening help? What treatments are available now and what might be possible in the future? These are just some of the questions that need answers if Nathan and Claire are to make an informed choice.

▲ **Figure 2.1** Should we or should we not?

Although the symptoms of cystic fibrosis and how it is inherited have been known for a long time – it was first recognised over 400 years ago – the genetic cause of the disease was only identified in the 1990s.

Overview of the biological principles covered in this topic

In this topic you will study how changes in DNA can result in genetic disease using the example of cystic fibrosis. You will first look in detail at the symptoms of cystic fibrosis, extending your GCSE knowledge of the fine structure of the lungs and gas exchange to see the importance of surface area to volume ratios in biology.

To understand the symptoms of cystic fibrosis you will study cell membrane structure and how substances move across membranes. To explain how faults arise in the cell membranes of a person with CF you will gain a detailed knowledge of how genes code for proteins, how proteins are made and how their function is dependent on their structure. You will find out how genes play their role in inheritance.

You will discover how cystic fibrosis is treated conventionally and how genetic screening and gene therapy may be used to help people with cystic fibrosis. Throughout the topic we'll consider the ethical issues raised by these new technologies and you will learn how ethical arguments can be evaluated.

Activity

Perform **Activity 2.1** to get an overview of cystic fibrosis and find out more about Claire's family and the problems she and Nathan could face in the future. **AS02ACT01**

Activity

Read the stories in **Activity 2.2** to see how cystic fibrosis has affected some people, and how they and their families cope. **AS02ACT02**

Review test

Are you ready to tackle topic 2 - *Genes and health*?

Complete the GCSE review before you start. **AS02RVT01**

2.1 The effects of CF

Cystic fibrosis one of is the most common genetic diseases in the UK, affecting approximately 7500 people. Every week five babies are born with CF and three young people die from cystic fibrosis, usually as a result of lung damage. In the 1960s the average life expectancy for a child with CF was just 5 years. By the year 2000 life expectancy had risen to 31. (To find out what it is today visit the Cystic Fibrosis Trust web site.)

Claire is well aware of the outward symptoms of CF; she sees the problems her Mum, Valerie, has with breathing, including a troublesome cough and repeated chest infections. Valerie also has to be very careful about her diet to ensure she overcomes problems with digestion and maintains her weight. But Claire and Nathan want to know in more detail what is going on within the gas exchange and digestive systems to cause these outward symptoms. Basically the symptoms are because Valerie has a sticky mucus layer lining many of the tubes and ducts in both systems.

How does cystic fibrosis affect breathing and gas exchange?

There is nothing unusual about having a layer of mucus in the tubes of the gas exchange system - everyone normally has a thin coating of mucus in these tubes. Any dust, debris or microorganisms that enter the airways become trapped in the mucus and are removed by the wave-like beating of cilia that line the tubes of the gas exchange system (Figure 2.2). Valerie has mucus that is drier than usual resulting in a sticky mucus layer that the cilia find difficult to move.

This sticky mucus in the lungs has two major effects on Valerie's health. It makes gas exchange less efficient and it increases the chances of lung infection.

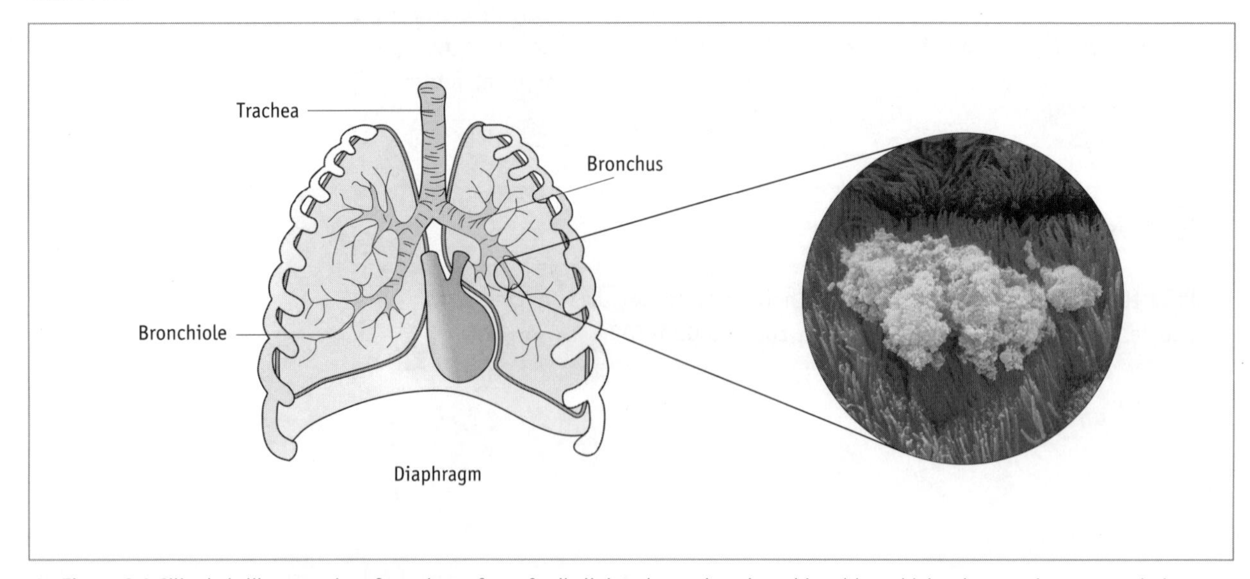

▲ **Figure 2.2** Cilia, hair-like extensions from the surface of cells lining the trachea, bronchi and bronchioles, have an important role in keeping the lungs clean. By a repeated beating motion they move mucus and particles up and out of the lungs. Scanning electron micrograph of cilia and mucus in the lungs of someone with CF.

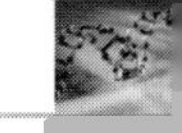

How the mucus makes gas exchange less efficient

Gases such as oxygen cross the wall of the alveoli into the blood system by **diffusion**.

Activity

If you have not observed diffusion in action **Activity 2.3** gives a practical demonstration. **AS02ACT03**

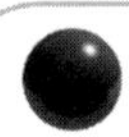

Key Biological Principle: Why are the alveoli so important?

Every living organism, from the tiniest single-celled creature to the largest mammal, has to take in materials and get rid of waste. The amount of material that has to be exchanged depends on the size of the organism. The larger the organism the more food and oxygen it will need and the more waste it will need to get rid of. In unicellular organisms the whole cell surface membrane is the exchange surface and oxygen and food diffuse into the cell down a **concentration gradient** (from high to low concentration). The gradient is maintained by the cell using the substances it absorbs.

Activity

Complete **Activity 2.4** to investigate the effect of surface area to volume ratio on uptake by diffusion. **AS02ACT04**

To understand about the absorption of substances in multicellular organisms you need to think about the size of the organisms surface area compared to its volume, this is known as the **surface area to volume ratio**. It is calculated by dividing the total surface area by the volume.

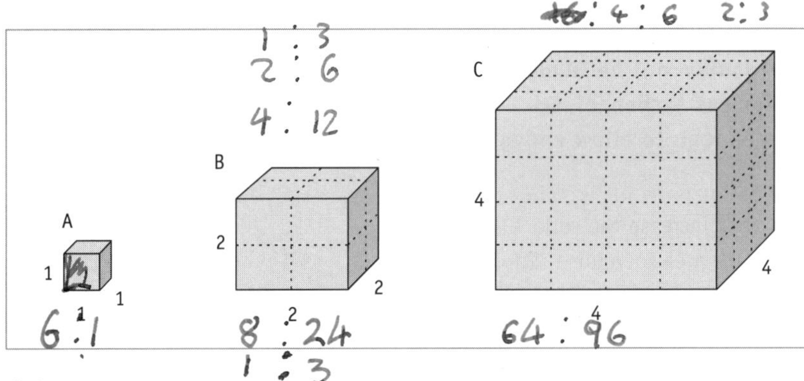

Q2.1 For each of the above 'organisms' work out its surface area, volume and then surface area to volume ratio.

Q2.2 As the organism grows larger what happens, quantitatively, to (a) its surface area, (b) its volume and (c) its surface area to volume ratio?

Q2.3 Assuming that this organism relies on diffusion across its outer surface for exchange, why would it have problems if it grew any larger?

Q2.4 If you compared a tiger, a horse and a hippopotamus, which would have the smallest surface area to volume ratio?

It is clear that as organisms get larger, the amount of surface area per unit of volume gets less and if they relied on this surface area for uptake of substances they would face a major problem absorbing enough to survive. How can an organism's volume increase while it still manages to absorb enough nutrients by diffusion?

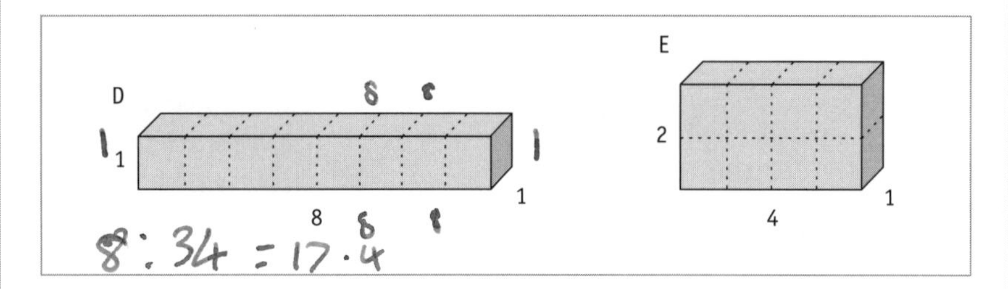

Q2.5 Work out the surface area, volume and surface area to volume ratio of organisms D and E.

Q2.6 Look at the values you have calculated for B, D and E what do you notice?

Q2.7 If a slug, an earthworm and a tape worm all had the same volume which would have the largest surface area to volume ratio?

All three blocks, B, D and E, have the same volume but very different surface areas. The most elongated block, D, has the largest surface to volume ratio. Relying on the outer body surface for gas exchange is only possible in animals with a very small volume or with a tubular or flattened shape, like worms.

Q2.8 Why would a land-living organism not be likely to use its entire external surface covering for gas exchange even when it had a surface area to volume ratio high enough to allow enough gas exchange to occur?

Larger organisms have special organs that increase the surface area for exchange, thus maintaining a high surface area to volume ratio.

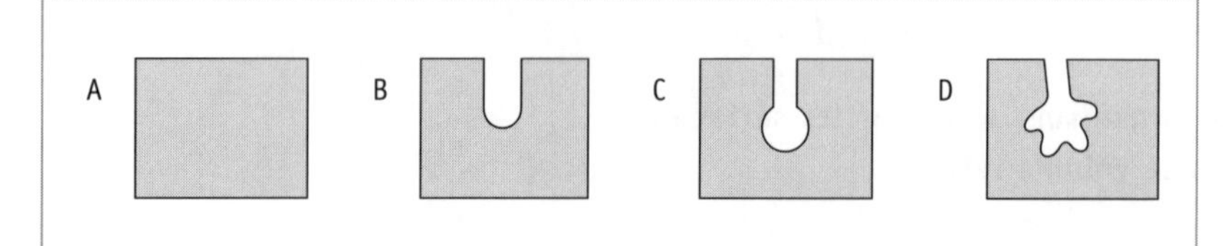

Q2.9 Which of these four organisms, all with the same volume, has the largest surface area for exchange?

Q2.10 Name two organs that have developed to aid exchange.

Q2.11 In humans, how do the substances that are absorbed get to all the distant parts of the body?

Within the lungs alveoli provide a large surface area for exchange of gases between the air and the blood. Look at Figure 2.3 and identify four features of the **gas exchange surface** that you think would ensure quick and efficient exchange between air in the alveoli and the blood.

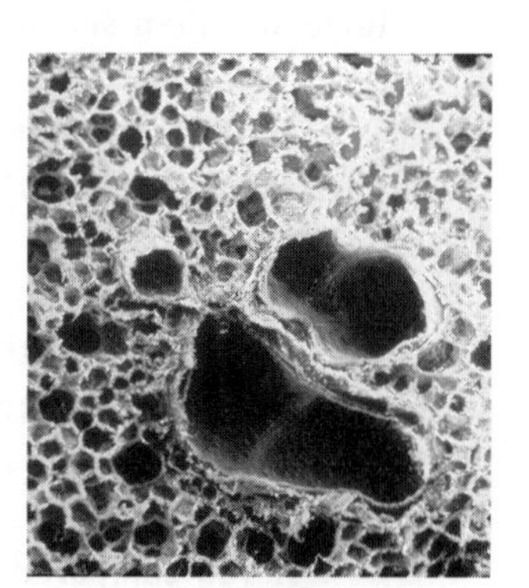

▲ **Figure 2.3** Ventilation of the lungs ensures that the air in the alveoli is frequently refreshed. This helps maximise gas exchange across the alveoli walls.

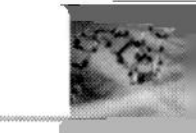

In **Activity 2.5** you will examine slides of the alveoli to observe the features that aid diffusion into the bloodstream. In the web tutorial, **Activity 2.6**, you can investigate the surface area of the lungs. **AS02ACT05, AS02ACT06**

The body's demand for oxygen is enormous so diffusion across the alveolar wall needs to be rapid. The rate of diffusion is dependent on three things: surface area, concentration gradient and thickness of the gas exchange surface. The rate of diffusion is directly proportional to the surface area and to the difference in concentration so that as the surface area increases the rate of diffusion increases while the greater the concentration gradient the faster the diffusion. The rate of diffusion is *inversely* proportional to the thickness of the gas exchange surface. The thicker the surface the slower the diffusion. These three factors can be combined as:

$$\text{Rate of diffusion} \propto \frac{\text{Surface area} \times \text{difference in concentration}}{\text{Thickness of the gas exchange}}$$

This is known as **Fick's Law**.

The sticky mucus layer in the bronchioles of a person with cystic fibrosis tends to block these narrow airways, preventing the ventilation of the alveoli below the blockage. This will reduce the number of alveoli providing surface area for gas exchange. Blockages are more likely at the narrow ends of the airways. These blockages will often allow air to pass when the person breathes in but not when they breathe out, resulting in overinflation of the lung tissue beyond the block. This can damage the elasticity of the lungs.

Any accumulation of the sticky mucus within individual alveoli will also reduce the surface area over which gases can be exchanged *and* increase the thickness of the gas exchange surface. Together these reduce the rate of diffusion.

Valerie finds it difficult to take part in physical exercise because her gas exchange system cannot deliver enough oxygen to her muscle cells. The oxygen is required to generate energy in the chemical processes of **aerobic respiration** and this energy is used to drive the contraction of the muscles as Valerie exercises. She becomes short of breath when taking exercise.

How mucus increases the chances of lung infections

The mucus in the gas exchange system can trap microorganisms, some of which can cause illness, i.e. are **pathogens**. The mucus is normally moved into the back of the mouth cavity where it is either coughed out or swallowed. With CF the mucus layer is so sticky that cilia cannot move the mucus.

Q2.12 Why might failure to move the mucus create such a problem?

Q2.13 Why does swallowing the mucus reduce the risk of infection?

Normally the body secretes naturally produced antibiotics into the mucus that help combat any infection. Unfortunately with cystic fibrosis the mucus contains high levels of salt which inactivates these antibiotics. The repeated infections can eventually weaken the body's ability to fight the pathogens, and cause damage to the structures of the gas exchange system. White blood cells fight the infections within the mucus but as they die they break down releasing DNA which increases the stickiness of the mucus.

How does cystic fibrosis affect digestion?

Most of the chemical breakdown of food molecules and the subsequent absorption of the soluble products into the bloodstream occurs in the small intestine. Glands secrete digestive enzymes into the lumen of the gut where they act as catalysts to speed up the breakdown. A wide range of enzymes are produced by glands in the gut wall and by exocrine glands outside of the gut, e.g. salivary glands, liver and the pancreas. Glands are **exocrine** if they secrete into a duct, unlike the **endocrine** glands which release hormones directly into the bloodstream. 85–90% of patients diagnosed with CF have problems with the exocrine function of the pancreas.

Groups of pancreatic cells produce enzymes that help in the breakdown of proteins, carbohydrates and lipids. These digestive enzymes are delivered to the gut in pancreatic juice released through the pancreatic duct (Figure 2.4).

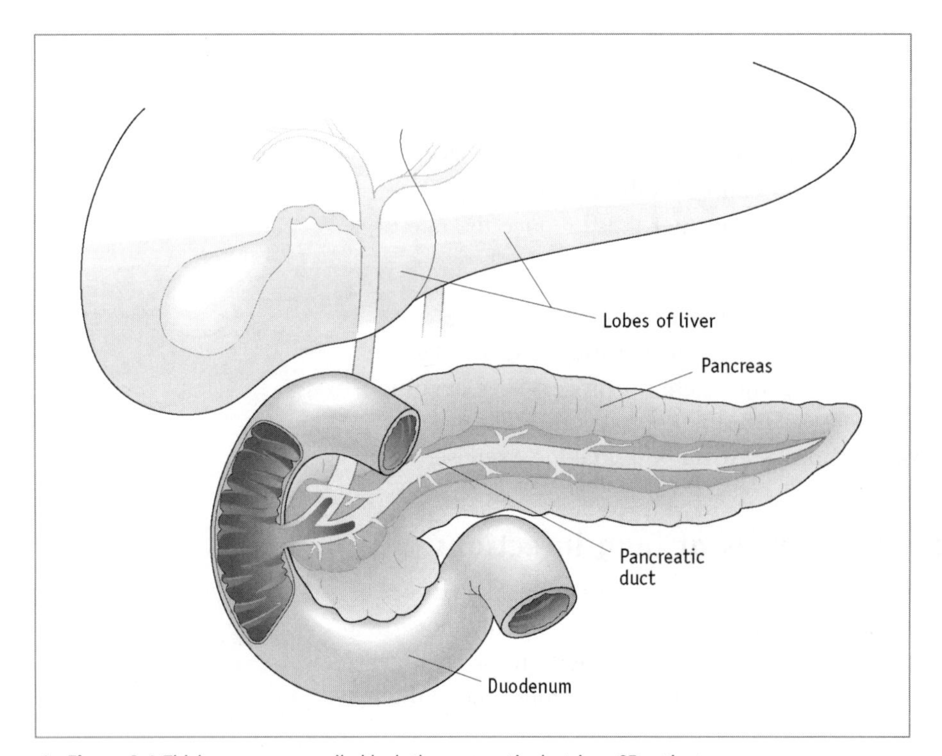

▲ **Figure 2.4** Thick mucus can easily block the pancreatic duct in a CF patient.

In a person with CF, the pancreatic duct becomes blocked by sticky mucus resulting in reduced release of digestive enzymes. The lower concentration of enzymes within the small intestine reduces the rate of digestion. Food is not fully digested and therefore not all the nutrients can be absorbed. This is called 'malabsorption syndrome'. Any molecules that are successfully broken down small enough to be absorbed have to pass across a thick sticky mucus layer to reach the blood system. This makes the absorption of food inefficient.

CF sufferers will have difficulty maintaining body mass because of the problems with digestion and absorption of nutrients. They have to eat more and include food of a high-energy content to ensure they obtain sufficient nutrients and energy. They require 120–140% of the recommended daily energy intake. People with CF may also take food supplements that contain digestive enzymes. The aim of these supplements is to help break down the food molecules.

An additional complication occurs when the pancreatic enzymes become trapped behind the mucus block in the pancreatic duct. These enzymes damage the pancreas itself. If the cells within the pancreas which produce the hormone insulin, involved in the control of blood sugar levels, become damaged this can lead to diabetes.

Activity

In **Activity 2.7** you investigate the effect of enzyme concentration on enzyme activity. **AS02ACT07**

Why is the mucus produced so sticky?

In people with CF, the mucus layer on the surface of the epithelial cells is sticky because it contains less water. The reduced water level is due to the reduction in the amount of chloride ions and water passing out of the epithelial cells. To understand what is going on you need to be clear about the structure of cell membranes and how substances are transported in and out of cells. For a cell to function correctly it needs to control transport across the cell membrane.

What are epithelial cells?

Epithelial cells form the outer surface of many animals including mammals. They also line the cavities and tubes within the animal, and cover the surface of internal organs. The cells work together as a tissue known as **epithelium**.

The epithelium may consist of one or more layers of cells sitting on a **basement membrane**. This is made of protein fibres in a jelly-like matrix. There are several different types of epithelia.

The walls of the alveoli and capillaries are **squamous** or **pavement epithelium**. The very thin flattened cells fit together like crazy paving. The cells can be less than 0.2 μm thick.

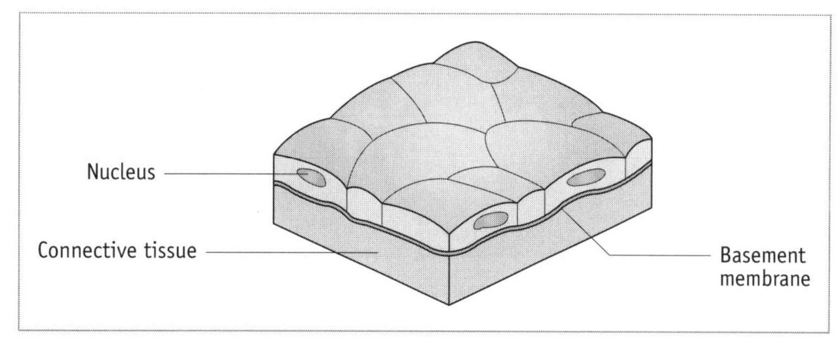

▲ Squamous or pavement epithelium

In the small intestine the epithelial cells extend out from the basement membrane. The column shaped cells make up **columnar epithelia**. The free surface facing the intestine lumen is normally covered in **microvilli** to increase surface area.

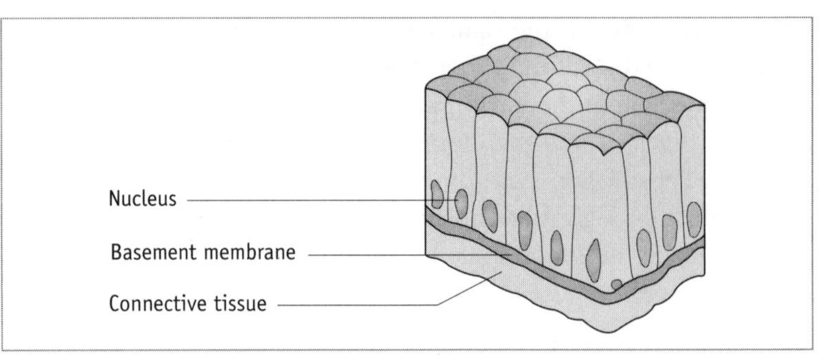

▲ Columnar epithelium

In the trachea, bronchi and bronchioles there are ciliated epithelial cells with cilia (hair-like structures) on the free surface. These cilia beat and move substances along the tube they line. The **ciliated columnar epithelium** of the gas exchange airways appears to be stratified (composed of several layers), but in fact each cell is in contact with the basement membrane. It is therefore known as **pseudostratified**.

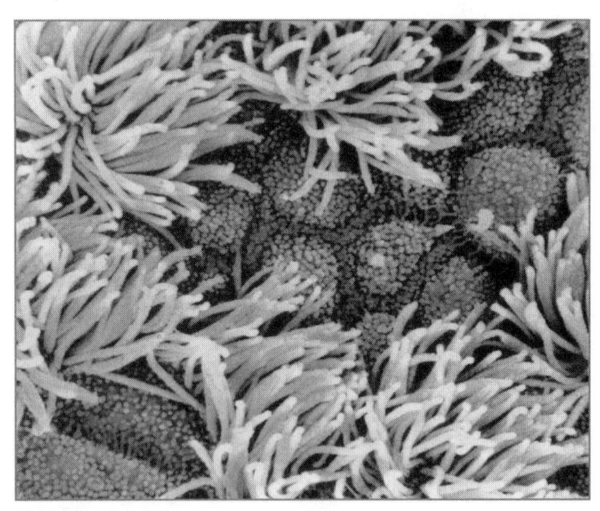

▲ Ciliated epithelium

Cell membrane structure

Under a light microscope the cell membrane looks like a single line. But closer examination using an electron microscope reveals that it is in fact a **bilayer**, appearing as two distinct lines about 7 nm wide (Figure 2.5). What do we mean by a bilayer? The basic structure is two layers of **phosopholipids**.

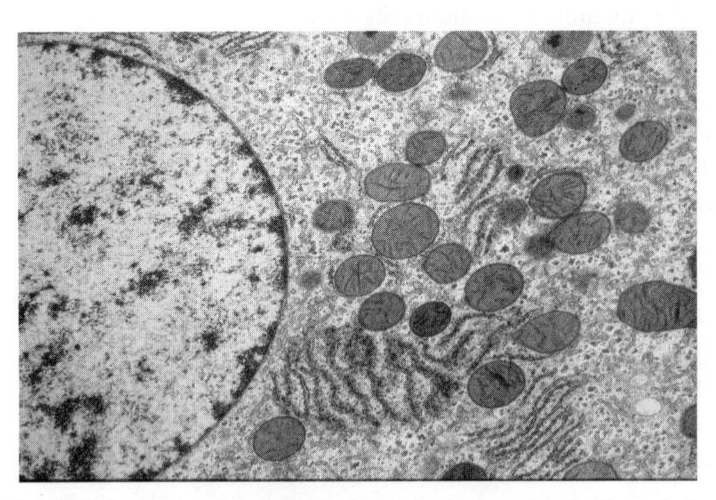

▲ **Figure 2.5** The cell membrane appears as two distinct layers when viewed with an electron microscope. Can you see any membranes within the cell that have a similar structure? The same type of membrane surrounds many cell organelles.

Look back at Figure 1.35 (page 35) and remind yourself of the structure of a lipid molecule – three fatty acids and a glycerol. Compare this to Figure 2.6 and notice the difference. In a phospholipid there are only two fatty acids; a phosphate group replaces one of the fatty acids

▲ **Figure 2.6** The phospholipid molecule has two distinct sections: the hydrophilic head and the hydrophobic tails. X can be a variety of chemical groups.

The phosphate head of the molecule is **polar**, which means that the sharing of the electrons within this part of the molecule is not quite even; one end becomes slightly positive and the rest is slightly negative. This makes the phosphate head attract other polar molecules, like water, and it is therefore **hydrophilic** (water-attracting). The fatty acid tails are non-polar and therefore **hydrophobic** (water-repelling). When added to water the phospholipids either form a layer on the surface with their hydrophobic tails directed out of the water or arrange themselves into clusters called micelles to avoid contact between the hydrophobic tails and water (Figure 2.7).

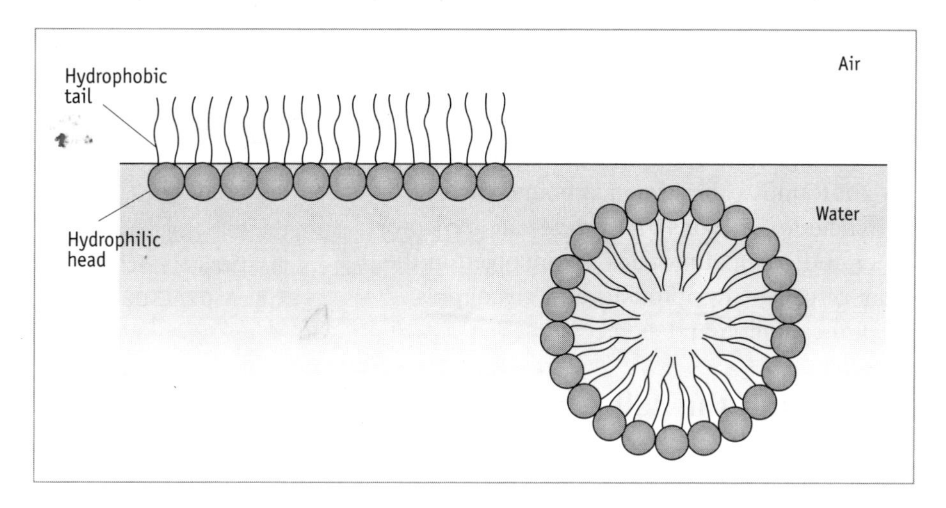

▲ **Figure 2.7** Phopholipids in water form a monolayer on the surface or micelles.

The phospholipds that make up the cell membrane form a bilayer with all the hydrophilic heads pointing outwards and the hydrophobic tails facing inwards avoiding any contact with water on either side of the membrane (Figure 2.8).

The cell membrane is not simply a phospholipid bilayer. It also contains proteins, cholesterol, **glycoproteins** (protein molecules with a polysaccharide attached) and **glycolipids** (lipid molecules with polysaccharides attached). Some of the proteins span the membrane. Other proteins are only found within the inner layer and some only within the outer layer (Figure 2.8). Membrane proteins have hydrophobic areas and these are positioned within the membrane bilayer.

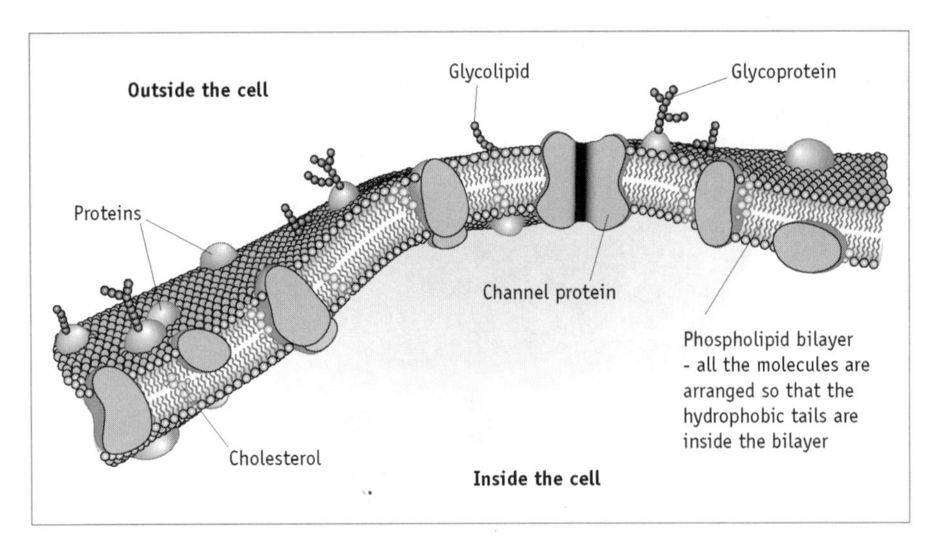

▲ **Figure 2.8** Diagram of the fluid mosaic model of the cell membrane.

It is thought that some of the proteins are fixed within the membrane but others are not and can move around in the fluid phospholipid layer. This arrangement is known as the **fluid mosaic model** of membrane structure. Singer and Nicolson first proposed it in 1972.

The more phospholipids containing unsaturated fatty acids there are present in the membrane the more fluid it is. Cholesterol also helps to maintain the fluidity of the membrane.

Q2.14 Can you suggest why the membrane is more fluid with unsaturated rather than saturated phospholipids making up the bilayer?

Activity

You can investigate the cell membrane practically in **Activity 2.8. AS02ACT08**

Many different types of proteins are found within the membrane, each type having a specific function. These include functions as enzymes and receptor molecules. As we shall see, carrier and channel proteins are involved in the transport of substances in and out of cells. Glycoproteins and glycolipids have important roles in cell-to-cell recognition and as receptors.

How do substances pass through the cell membrane?

Molecules and ions can move across membranes by **diffusion**, **osmosis** or **active transport**. Bulk transport of substances is by **exocytosis** and **endocytosis** (Figure 2.9). See page 44 for an example of exocytosis: neurotransmitters are released from the cell as a vesicle fuses with the membrane. Endocytosis is the reverse process: substances are taken into a cell by the creation of a vesicle. Part of the cell membrane engulfs the solid or liquid material to be transported.

Diffusion

Diffusion is the movement of molecules or ions from an area of their high concentration to an area of their low concentration. Diffusion will continue until the substance is evenly spread throughout the whole volume. Small molecules like oxygen and carbon dioxide can diffuse rapidly across the cell membrane. Carbon dioxide is polar but its small size allows rapid diffusion.

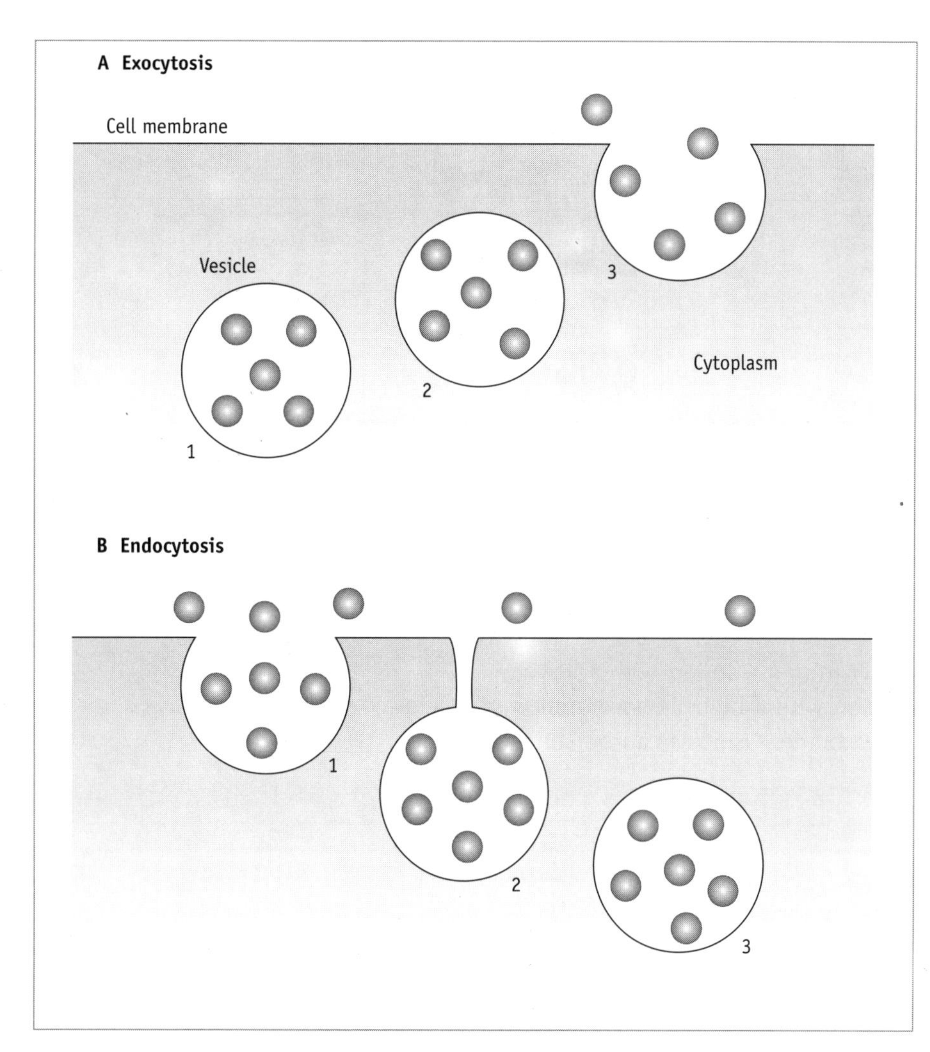

Figure 2.9 In exocytosis, vesicles fuse with the cell membrane, releasing their contents from the cell. In endocytosis, substances are brought inside a cell within vesicles formed from the cell membrane.

Hydrophilic molecules and ions any larger than carbon dioxide cannot simply diffuse through the bilayer because the hydrophobic tails of the phospholipids provide an impenetrable barrier to them. Instead they cross the membrane with the aid of proteins in a process called **facilitated diffusion**. They may diffuse through water-filled pores within **channel proteins** that span the membrane (Figure 2.8). There are different channel proteins to allow the passage of different molecules. Each type of channel protein has a specific shape that only permits the passage of one particular molecule. Some of the channels can be opened or closed depending on the presence or absence of a **signal** which could be a specific molecule, for example a hormone, or a change in potential difference (voltage across the membrane).

Some of the proteins that play a role in facilitated diffusion are not just simple channels but are **carrier proteins**. The ion or molecule binds onto a specific site on the protein. The protein changes shape (Figure 2.10) and as a result the ion or molecule crosses the membrane. The movement can occur in either direction, with the net movement being dependent on the concentration difference across the membrane. Molecules move from high to low concentration due to more frequent binding to carrier proteins on the side of the membrane where the concentration is higher.

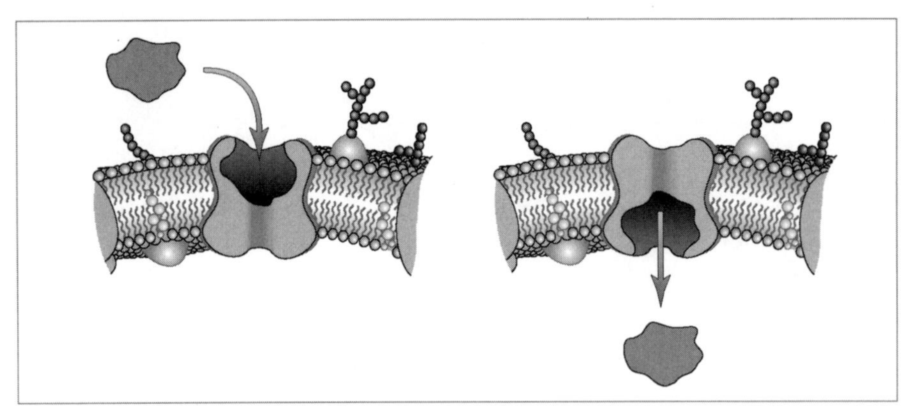

▲ **Figure 2.10** The carrier protein changes shape, facilitating diffusion.

Diffusion, whether facilitated or not, is sometimes called **passive transport**. 'Passive' here refers to the fact that no energy is expended in the movement.

Osmosis

Osmosis is the movement of water from a solution with a lower concentration of solute to a solution with a higher concentration of solute through a partially permeable membrane. Osmosis can be summarised as follows:

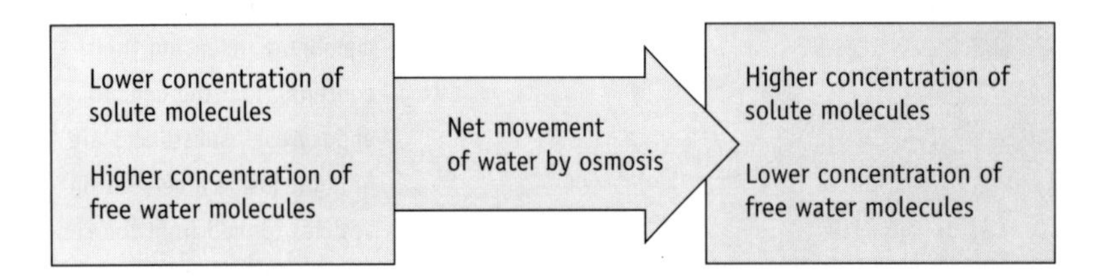

Look at the upper and lower parts of Figure 2.11. In each case decide the direction in which the solvent (i.e. water) molecules will have a net (overall) tendency to move. Only the lower half of Figure 2.11 has a partially permeable membrane so only here can osmosis take place. Osmosis is due to the random movement of water molecules across the membrane and is a particular type of diffusion. If solute molecules are present water molecules form weak hydrogen bonds with these solute molecules which reduces their movement. The more solute present the more water is associated with the solute and less water is free to collide with and move across the membrane.

Active transport

If substances need to be moved across a membrane against a concentration gradient (from low concentration to high concentration) then energy is required. As with facilitated diffusion, specific carrier proteins are involved. The energy comes from respiration and is supplied by the energy transfer molecule **ATP**. The substance to be transported binds to the carrier protein. The energy is used to bring about the change in shape of the carrier protein and the substance is released on the other side of the membrane.

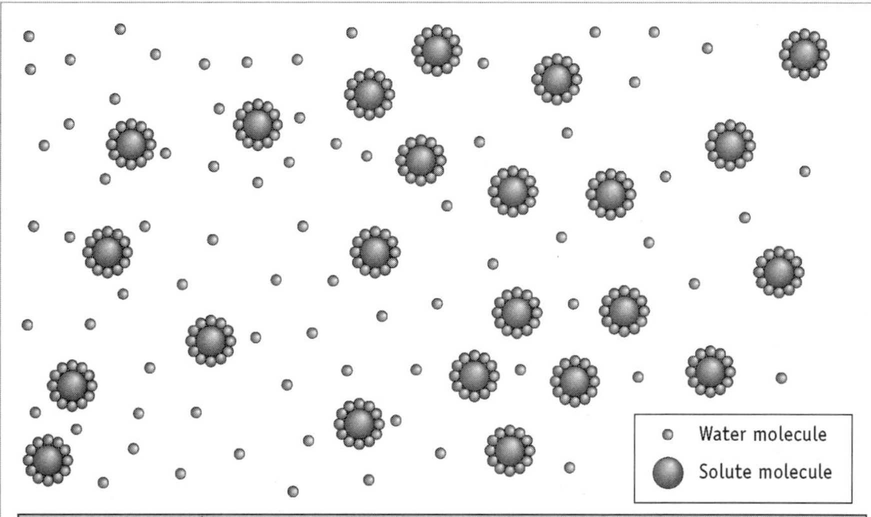

With unrestricted movement of the molecules there will be diffusion of both solute and water in both directions until equilibrium is achieved.

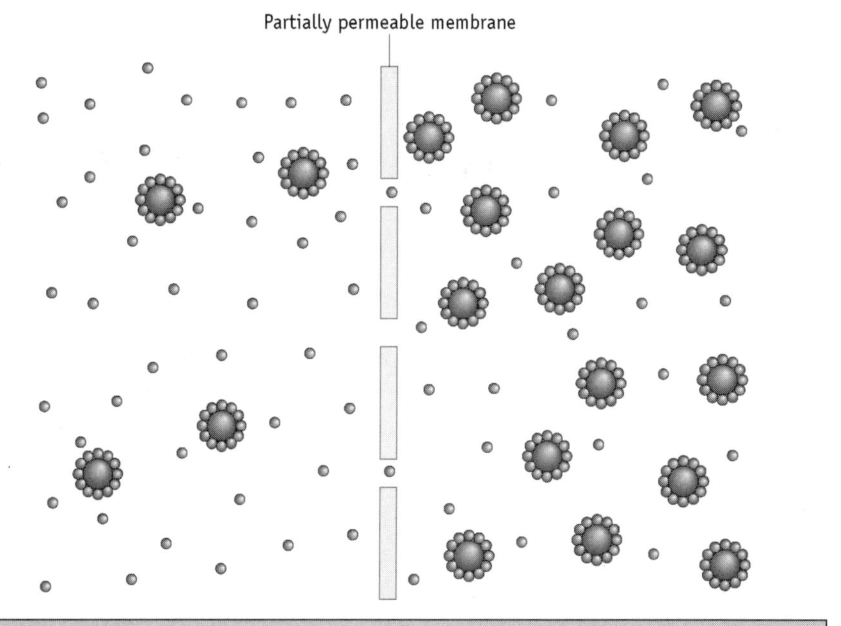

The presence of a partially permeable membrane prevents the movement of some molecules. The solute cannot cross the membrane but free water molecules (those not associated with solute molecules) can diffuse through the membrane to achieve equilibrium.

◀ **Figure 2.11** In which direction will the water molecules move?

Q2.15 For each of these examples suggest the type of transport most likely to be involved:

a) movement of oxygen across the wall of an alveolus

b) absorption of phosphate ions into root hair cells

c) pumping of calcium ions into storage vesicles (small, storage membrane-bound sacs) inside muscle cells

d) release of glucose from liver cells into the bloodstream

e) removal of the sodium ions that diffuse into the nerve cell, thus maintaining a low concentration within the nerve axon

f) reabsorption of water molecule from the kidney tubule.

What is happening within the cells lining the airways?

In people who do not have cystic fibrosis, chloride ions are actively pumped into the epithelial cells of the airways (and elswhere) by chloride pumps located in the cell membrane on the tissue side of the epithelium (Figure 2.12). The concentration of chloride ions within the cells rises, creating a concentration gradient. The chloride ions diffuse out of the cells through chloride channels located in the membrane facing the airway lumen. This channel protein is known as the cystic fibrosis transmembrane regulatory (**CFTR**) channel protein. It is a **gated** protein channel that is *only* open when it is associated with ATP.

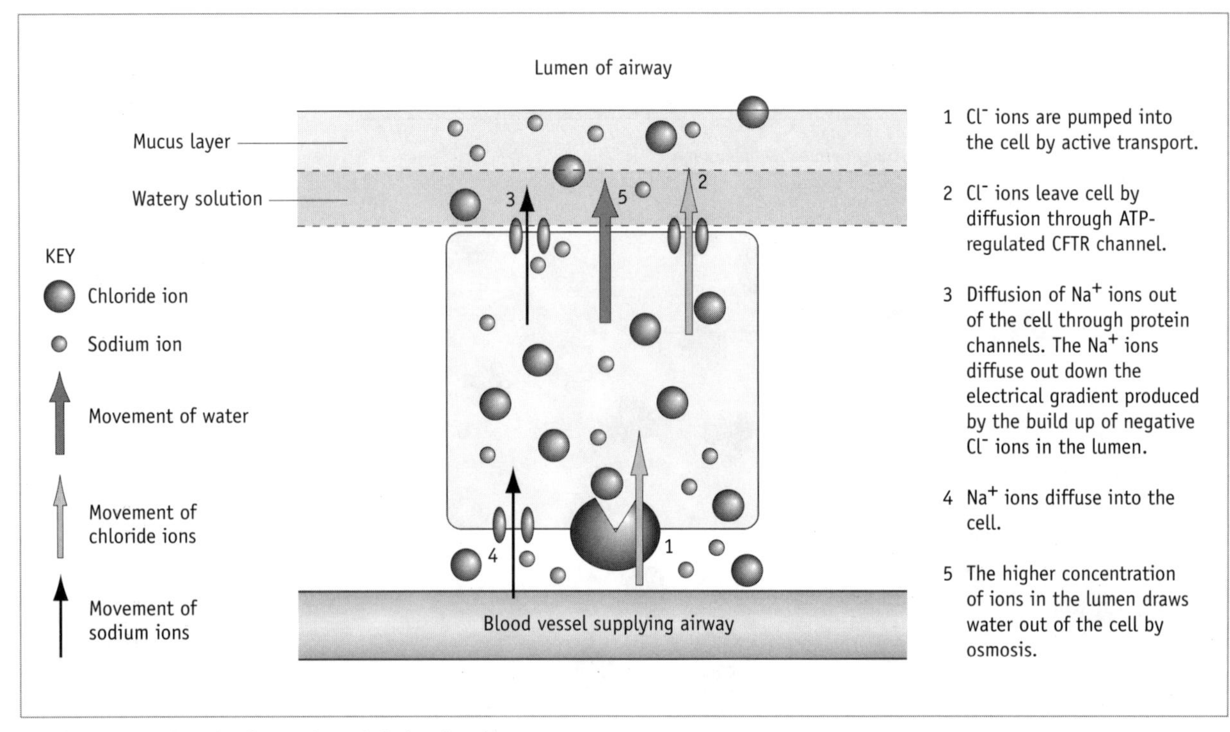

▲ **Figure 2.12** The role of CTFR channels in keeping the mucus runny.

The build up of negatively charged chloride ions (Cl^-) in the fluid that lines the airway creates an electrical gradient across the membrane. Sodium ions (Na^+) diffuse out of the cell down the electrical gradient. The sodium ions will continue to move out until there is an approximate balance in electrical charge across the membrane. So the sodium ion concentration within the cell falls and sodium ions diffuse *into* the cell through sodium channels on the *tissue* side of the cell.

The secretion of sodium and chloride ions into the lumen of the airways draws water out of the cell by osmosis. This water keeps the mucus lining the airway runny.

In a person who has CF the CFTR protein may be missing or, if present, it does not function properly. It does not allow the chloride ions that are being pumped into the cell to leave. Chloride ion concentration in the cell builds

up. The high solute concentration in the cell causes water to move *into* the cell by osmosis. The mucus lining the epithelial cells of the airways therefore becomes thick and sticky.

Activity

Work through the interactive tutorial in **Activity 2.9** to investigate the effects of a functioning and a non-functioning CFTR protein channel on salt and water secretion in the airways. **AS02ACT09**

We have seen how sticky mucus produced by a defective CFTR protein can lead to major complications in the lungs and pancreas. In the reproductive system is can also cause severe problems. Females have a reduced chance of becoming pregnant because a mucus plug develops in the cervix. This stops the sperm from reaching the egg. Males with cystic fibrosis commonly lack the vas deferens (sperm duct) on both sides, which means sperm cannot leave the testis. Where the vas deferens is present it can become partially blocked by a thick sticky mucus layer. This results in a reduction in the number of sperm present in each ejaculate.

Q2.16 Salty sweat is often one of the first signs that a baby may have CF. Why might the sweat of a person with CF be more salty than normal? (Hint: the CFTR protein in sweat ducts works in the opposite direction.)

2.2 How is the CFTR protein made?

The CF gene, a section of DNA carrying the instructions to make the CFTR protein, is located on Chromosome 7. The protein it codes for is 1480 amino acids long and these are arranged into the three-dimension structure shown in Figure 2.13.

The defective CFTR protein is caused by an error in the DNA that codes for it. In order to know how DNA works (and what has gone wrong in cystic fibrosis) we need to know what a **gene** is and how the codes it contains are used to make proteins.

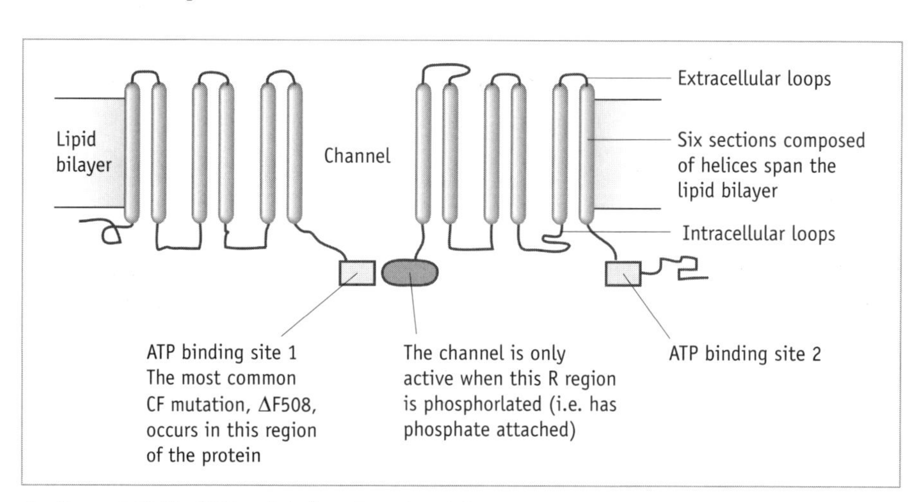

▲ **Figure 2.13** The CFTR protein is a channel protein.

Extension

You can read about the controversy surrounding the work of Rosalind Franklin and DNA in the extension material. **AS02EXT01**

A gene is a segment of DNA which codes for one protein. Each chromosome carries numerous genes, although they make up only a fraction of the total length of DNA. The job of the remainder is not fully known. Together all the genes in an individual (or species) are known as the **genome**.

Nice to know: DNA and immortality

'The seed of life itself. Peel the chains apart, each chain reproduces the other, one becomes two, two become one. Generation on generation, all the way from Adam and Eve to you and me. It never dies. One simple shape. The womb of humanity. Endlessly, effortlessly fertile, dividing, reforming . . . It's the closest we'll ever get to immortality.'

Tim Piggott-Smith who acted Francis Crick in the film "Life Story" talking about the structure of DNA

▲ **Figure 2.14** Watson (left) and Crick (right) worked out the structure of DNA in 1953.

In 1953 James Watson and Francis Crick proposed a model for the structure of DNA using the X-ray diffraction patterns of Rosalind Franklin and Maurice Wilkins. Their model was correct and their discovery has revolutionised biology. DNA is found in every cell nucleus. It contains the genetic code which dictates all our metabolism and inherited characteristics. It does this by controlling the manufacture of proteins. Your proteins are what make you unique; they are what make you a human being and not an oak tree or a chimpanzee. They play a vital role in giving you the unique characteristics that mean you are not the same as the person sitting next to you (you are probably rather glad about that).

How DNA codes for proteins

Deoxyribose nucleic acid (DNA) is a long chain molecule made of many units called **nucleotides**.

Q2.17 Why is DNA called a polynucleotide?

A nucleotide contains three molecules linked together by condensation reactions: **deoxyribose** (a 5-carbon sugar), a **phosphate group** and an **organic base** containing nitrogen. Look at Figure 2.15 to see how these three are arranged in a nucleotide.

Nucleotides link together by condensation reactions between the sugar of one nucleotide and the phosphate of the adjacent one producing a long chain of nucleotides. The nitrogen-containing base is the only part of the nucleotide that is variable. There are four bases – **adenine**, **cytosine**, **guanine** and **thymine**. Bases are frequently represented by their initial letter: **A**, **C**, **G** and **T** respectively.

In a DNA molecule there are two long strands of nucleotides twisted around each other to form a double helix (Figure 2.16), rather like a spiral staircase.

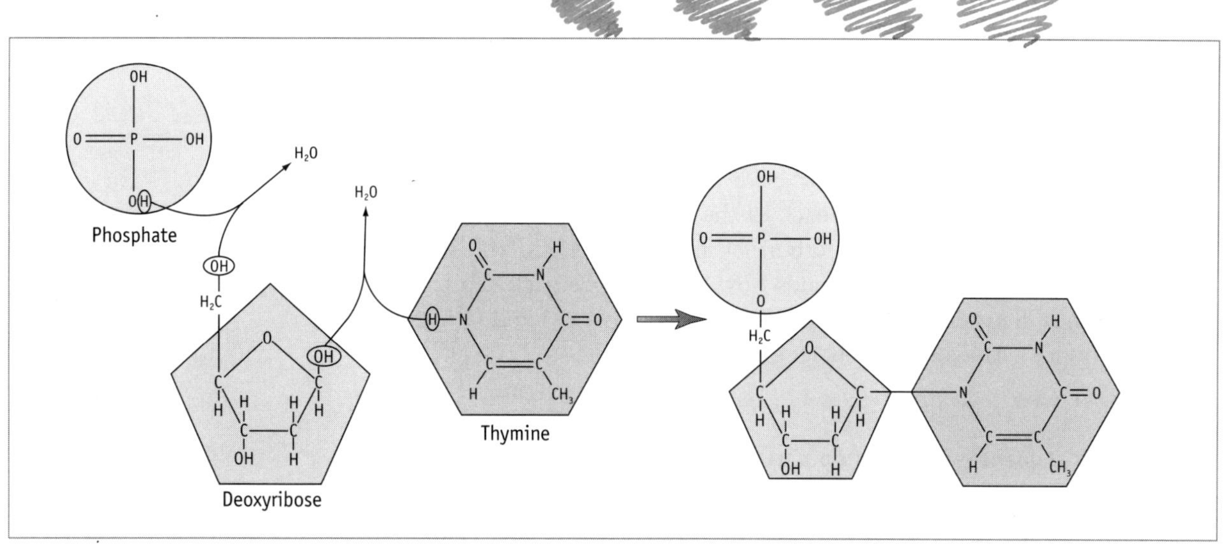

▲ **Figure 2.15** A phosphate, a deoxyribose sugar and a base join to form a nucleotide.

The sugars and phosphates form the 'backbone' of the molecule and are on the outside. The bases point inwards horizontally, rather like the rungs of a ladder. The two strands which run in opposite direction are known as antiparallel strands and are held together by weak hydrogen bonds between pairs of bases. The DNA in each human cell contains some 3000 million of these base pairs.

If you look at Figure 2.16 you should notice that the bases only pair in a certain way: adenine only pairs with thymine, and cytosine only pairs with guanine.

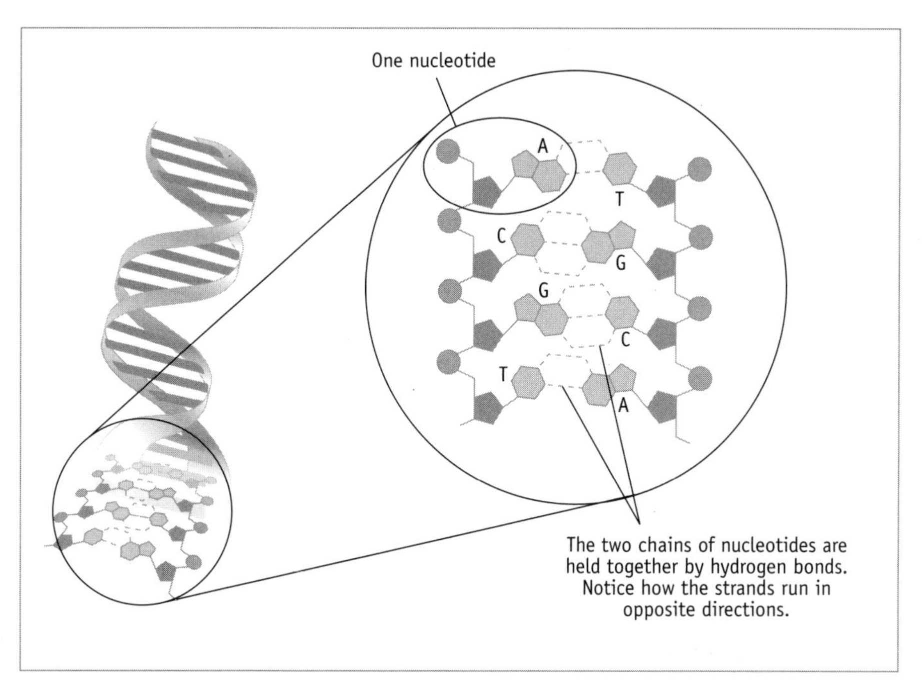

▲ **Figure 2.16** The structure of DNA. Remember that any picture that you see of DNA is a simplification, a model based on evidence from techniques such as X-ray diffraction. A whole DNA molecule is much longer than this and usually contains millions of base pairs.

Q2.18 Look at Figure 2.16 and suggest why the bases might only form these pairs.

The key is in the structure of the bases and the bonding. A and G both have a two ring structure whereas C and T have only one. The bases pair so that there are effectively three rings at each of the rungs of the DNA molecule, making the molecule a uniform width along its whole length. The shape and chemical structure of the bases dictates how many bonds each can form and this determines the pairing of A with T (2 hydrogen bonds) and C with G (3 hydrogen bonds). This deceptively simple fact is the clue to how DNA works.

Q2.19 The sequences of bases on part of one strand of a DNA molecule is:

A T C C C T G A G G T C A G T

What would be the sequence of bases on the corresponding part of the other strand?

Activity

Try making a 3D model of DNA in **Activity 2.10** and extract DNA from onions in **Activity 2.11. ASO2ACT10, ASO2ACT11**

DNA carries the genetic information passed from one generation to the next. It helps determine the structure and function of the cell by telling the cell what proteins to make. Protein molecules are strings of amino acids and, as we shall see, the structure and function of each different protein is determined by the sequence of the amino acids (page 75). The CF gene on chromosome 7 is a long one, made up of about 230 Kbp (230 000 base pairs) and instructs the cell to make the CFTR protein that forms the transmembrane chloride channel. But what is the code? How does the sequence of DNA bases tell the cell which amino acids to link together to make the CFTR or any other protein?

The triplet code

The genetic code is not simply one base coding for one amino acid. This would mean proteins could only contain 4 different amino acids whereas 20 amino acids are found commonly in proteins. The code carried by the DNA is a three base or **triplet code**. Each adjacent group of three bases codes for an amino acid; the triplets do not overlap. There are 64 three-letter combinations possible; some are start signals and others stop signals (called chain terminators). Several triplets code for the same amino acid. In some cases all the codes with the same first two letters code for the same amino acid. This amazingly simple but fundamental coding system is found in all organisms.

From DNA to proteins

But DNA cannot pass through the membranes surrounding the nucleus into the cytoplasm where proteins are constructed. How do the instructions get from the nucleus to where they are needed in the cytoplasm? It is rather like producing a photocopy of some very precious original plans that cannot leave head office so that everyone on the factory floor can work from the original. A 'copy' of the DNA is made. This 'copy' is not DNA but another type of nucleic acid called **RNA**, **ribonucleic acid**. This RNA leaves the nucleus carrying the information to the cytoplasm where it is used in the manufacture of proteins.

What is the difference between DNA and RNA?

An RNA molecule has a *single* strand made of a string of RNA nucleotides. These are very similar in structure to the DNA nucleotides except they contain **ribose** sugar and not deoxyribose (Figure 2.17). They also contain the base **uracil (U)** instead of thymine (so RNA *never* contains thymine).

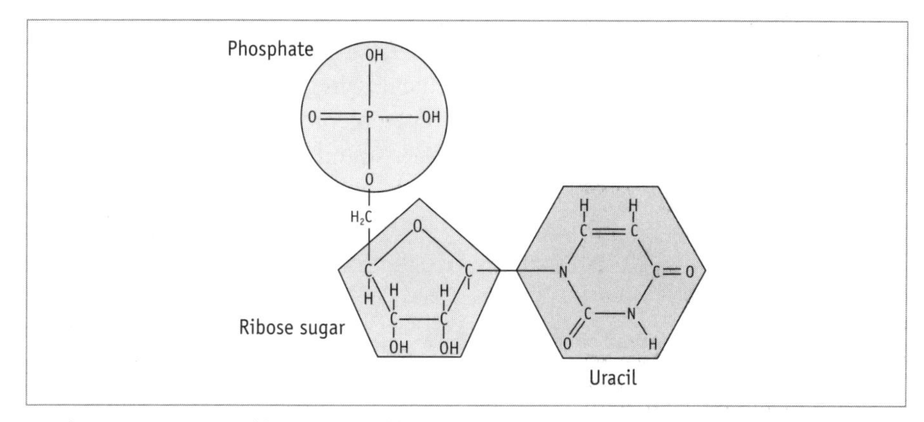

▲ **Figure 2.17** Compare this RNA nucleotide with the DNA nucleotide in Figure 2.14 and notice the differences.

There are three types of RNA involved in protein synthesis: **messenger RNA (mRNA)**, **transfer RNA (tRNA)** and **ribosomal RNA (rRNA)**. Their different roles in protein synthesis can be seen in Figure 2.18.

Protein synthesis has two stages; the first occurs in the nucleus and the second in the cytoplasm.

1 Transcription

Transcription takes place in the nucleus. The DNA double helix unwinds and the sequence on one of the strands, the **template strand**, is used in the production of a messenger RNA molecule. This mRNA is built from free RNA nucleotides which line up alongside the DNA strand. Because of base pairing the order of bases on the DNA exactly determines the order of the bases on the RNA. In other words, every triplet code on DNA gives rise to a complementary **codon** on messenger RNA. This complementary base pairing can be seen in Figure 2.18. This process involves a number of enzymes such as RNA polymerase. The completed messenger RNA molecule now leaves the nucleus through a pore in the **nuclear envelope** (the two membranes that surround the nucleus) and enters the cytoplasm. This is where the second stage takes place.

2 Translation

Translation can only start once the messenger RNA molecule becomes attached to a **ribosome**. Ribosomes are small organelles made of ribosomal RNA and protein. Ribosomes are found free in the cytoplasm or attached to **endoplasmic reticulum**, a system of flattened, membrane-bound sacs. A transfer RNA molecule also becomes attached, its **anticodon** bonding with the messenger RNA codon. Then the amino acid that the transfer RNA has carried becomes attached to a growing chain of amino acids by means of a **peptide bond**. The ribosome moves along the messenger RNA until all the codons have been used and the complete chain of amino acids has been produced. See Figure 2.18.

Activity

The computer simulation in **Activity 2.12** looks at nucleic acid structure and the sequence of events in protein synthesis. **AS02ACT12**

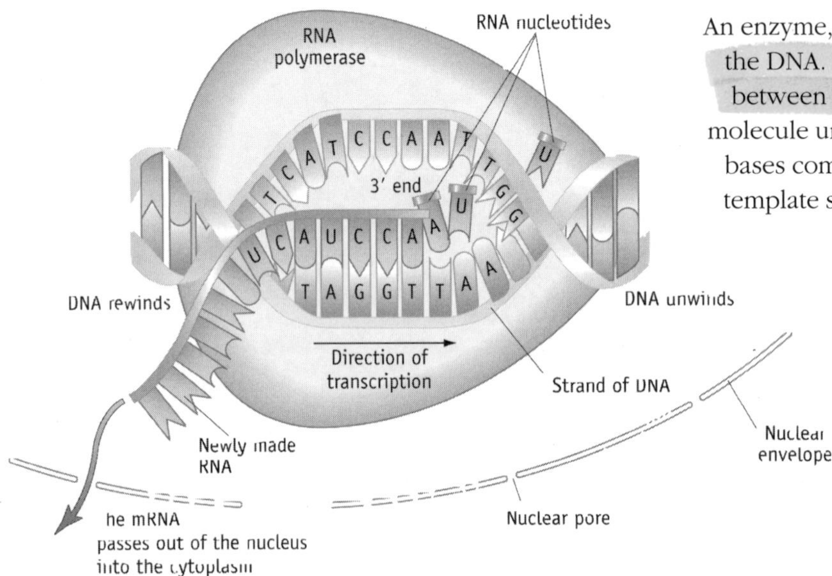

An enzyme, RNA polymerase, attaches to the DNA. The weak hydrogen bonds between bases break and the DNA molecule unwinds. RNA nucleotides with bases complementary to those on the template strand of the DNA pair up and bond to form a mRNA molecule. When complete, the mRNA molecule leaves the nucleus through a pore in the nuclear envelope. The template strand is also known as the antisense strand because once transcribed it makes a mRNA molecule with the same base sequence as the DNA coding strand. The coding strand is known as the sense strand.

▲▼ **Figure 2.18** Protein synthesis.

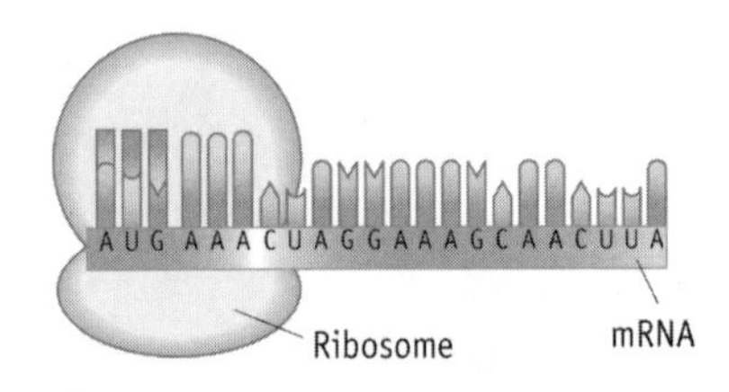

The mRNA attaches to the surface of a ribosome. Ribosomes are made up of two sub units. The mRNA attaches to the smaller sub unit so that two mRNA codons face the two binding sites of the larger sub unit.

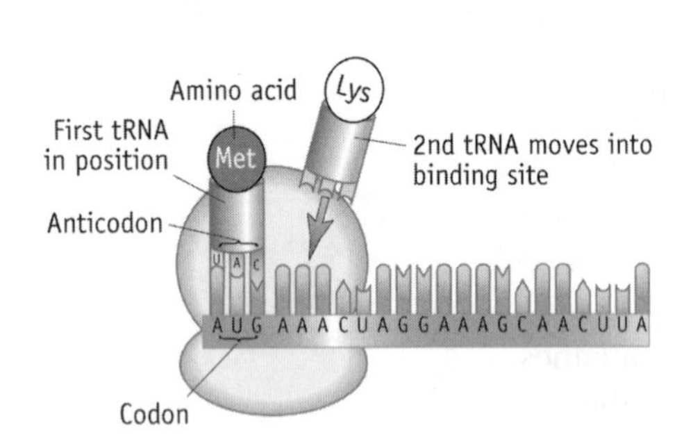

At one end of the tRNA is a triplet base sequence called an anticodon. The three bases of the anticodon are complementary to the mRNA codon for an amino acid. For example, the mRNA codons for the amino acid proline are CCC, CCA, CCG and CCU. The complementary anticodons are GGG, GGU, GGC, and GGA. Within the cytoplasm free amino acids become attached to the correct tRNA molecules. Each amino acid has its own specific tRNA which carries it to the ribosome.

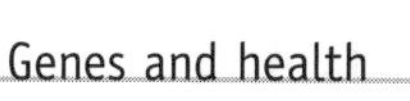

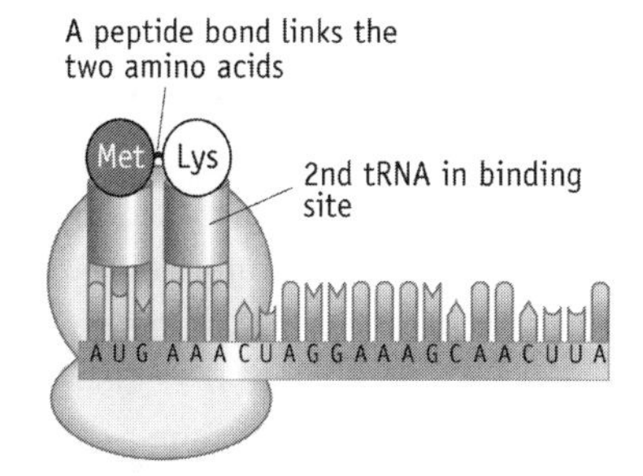

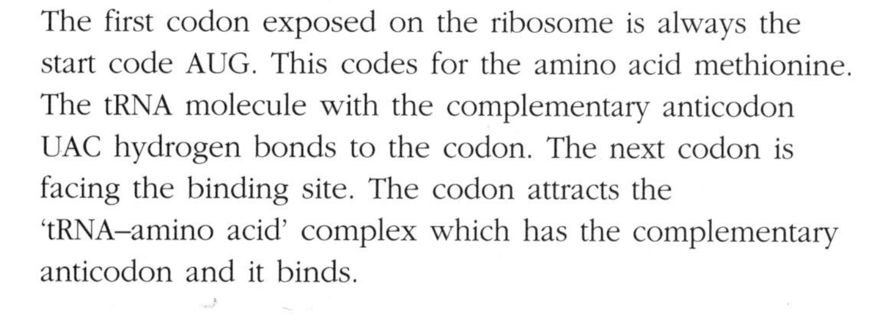

The first codon exposed on the ribosome is always the start code AUG. This codes for the amino acid methionine. The tRNA molecule with the complementary anticodon UAC hydrogen bonds to the codon. The next codon is facing the binding site. The codon attracts the 'tRNA–amino acid' complex which has the complementary anticodon and it binds.

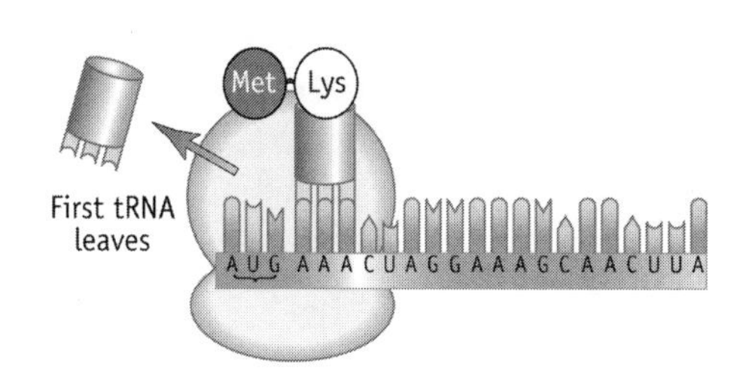

The ribosome holds the mRNA, tRNAs, amino acids and associated enzyme in place while a peptide bond forms between the two amino acids. The two amino acids are linked by a condensation reaction between the amine group of one peptide and the carboxylic acid group of the next forming a dipeptide.

Once the bond has formed the ribosme moves along the mRNA to reveal a new codon at the binding site. The first tRNA returns to the cytoplasm. The whole process is repeated and translation continues until the ribosome reaches a stop signal: UAA, UAC, or UGA.

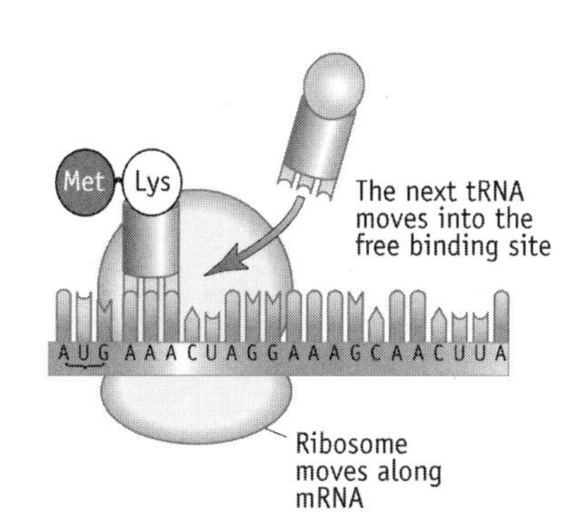

A protein molecule can be made up of thousands of the 20 naturally occurring amino acids. The sequence of the amino acids, its primary structure, determines the structure and properties of the protein and a slight variation in the order, for example changing even one of the amino acids in the chain, may substantially alter the protein's structure and properties.

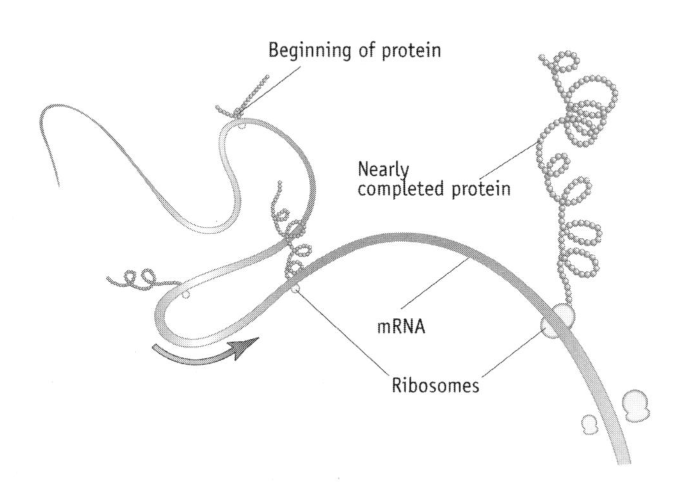

▲ **Figure 2.18 (cont.)** Protein synthesis.

If several ribosomes attach to a single mRNA molecule several copies of the same protein can be produced at the same time. As the protein molecule is formed, it folds up into the three-dimensional shape determined by its primary structure – the sequence of amino acids. The amino acids in different parts of the chain interact with each other forming bonds which hold the protein molecule in a precise shape. Some sections of the chain may coil into a helical structure or a pleated sheet, known as its secondary structure, before folding to form a specific three-dimensional shape known as its tertiary structure. The transmembrane sections of the CFTR protein (see Figure 2.22) are composed of helices.

Q2.20 What would be the sequence of bases on a length of messenger RNA built using the following DNA strand as a template?

T A C A T G G A T T C C G A T

Q2.21 How many tRNA molecules would be involved in the synthesis of the protein coded for by this section of DNA?

Q2.22 What are the anticodons, assuming you read the section from left to right?

Web link

The interactive tutorial in **Activity 2.14** lets you review how amino acids join to form a polypeptide and then fold to achieve their three dimensional structure. **ASO2ACT14**

Activity

You might like to try the protein synthesis cut and stick in **Activity 2.13** to check that you follow what is going on. **ASO2ACT13**

Key biological principle: Protein structure is the key to protein function

Proteins have a very wide range of functions within living organisms. It's the precise shape adopted by the protein as a result of its amino acid sequence that enables it to perform its specific function. Proteins can be divided into two distinct groups: globular and fibrous proteins.

In **globular proteins** the polypeptide chain is folded into a compact spherical shape which is held in place by bonding between amino acids. Enzymes are an important group of globular proteins. Their three-dimensional shape is crucial to their ability to form enzyme-substrate complexes and catalyse reactions within cells. Transport proteins within membranes, the oxygen transport pigments, haemoglobin in red blood cells and myoglobin (Figure 2.19) within muscle cells are also globular and rely on shape for binding. Antibodies too are globular and rely on their precise shape to bind to the countless microorganisms that assail us.

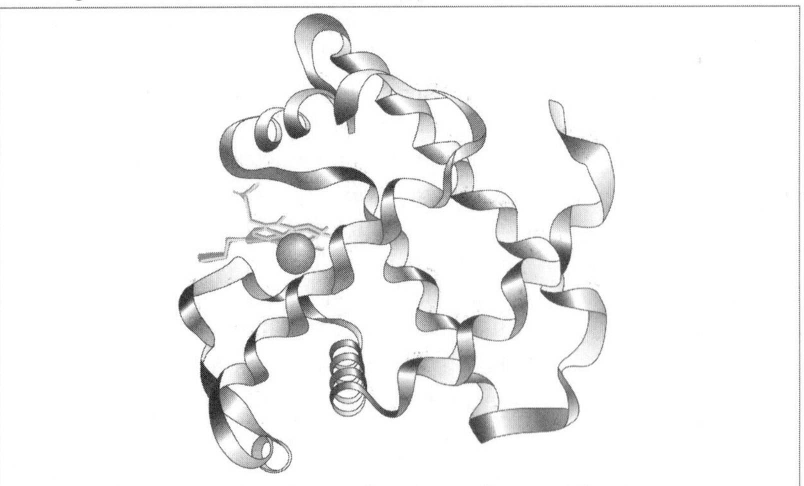

▲ **Figure 2.19** A globular protein, myoglobin, acts as an oxygen-storage molecule in muscle cells. The oxygen, shown as a circle attaches to the iron within the haem group. Because the protein is associated with another chemical group it is often referred to as a conjugated protein.

Fibrous proteins do not fold up into a ball shape; they remain as long chains often with several polypeptide chains cross-linked together for additional strength. These insoluble proteins are important structural molecules. **Keratin** in hair and skin, and **collagen** (Figure 2.20) in the skin, tendons, bones, cartilage and blood vessel walls are examples of fibrous proteins.

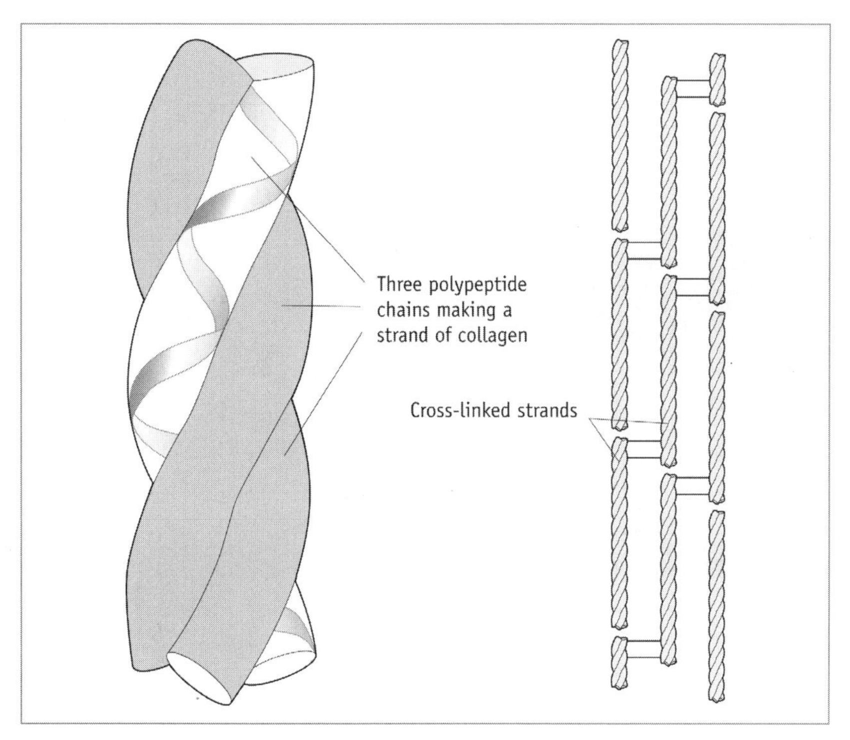

▲ **Figure 2.20** Collagen is a fibrous protein. Three polypeptide chains wind around each other to form a rope-like strand held together by hydrogen bonds between the chains. Each strand cross-links to other strands to produce a molecule with tremendous strength. Notice that the strands are staggered to avoid the creation of any weak points along the length of the molecule.

How enzyme function is dependent on three-dimensional structure

Enzymes act as **biological catalysts**. They speed up chemical reactions that would otherwise occur very slowly at the temperature within cells. The precise three-dimensional shape adopted by an enzyme includes a depression on the surface of the molecule called the **active site**. This site may be a relatively small part of the large protein molecule. Only a few amino acids may be directly involved in the active site with the remainder of the molecule maintaining the three-dimensional shape of the protein molecule (Figure 2.21).

Either a single molecule with a complementary shape, or more than one molecule that together have a complementary shape, can fit into the active site. These **substrate molecule(s)** form temporary bonds with the amino acids of the active site to produce an **enzyme-substrate complex**. The enzyme holds the substrate molecule(s) in such a way that they react more easily. When the reaction has taken place the **products** are released, leaving the enzyme unchanged. The substrate is often likened to a 'key' which fits into the enzyme's 'lock', so this is known as the **lock-and-key** theory of enzyme action.

Each enzyme will only catalyse one **specific** reaction because of its precisely shaped active site.

It has also been found that the active site is often flexible. When the substrate (or substrates) enters the active site, the molecule changes shape, fitting more closely around the substrate (Figure 2.21). It is like a person putting on a wet suit; the wet suit shape changes to fit the body but returns to its original shape when taken off. This is known as the **induced fit** theory of enzyme action. Only a specific substrate will induce the change in shape of an enzyme's active site.

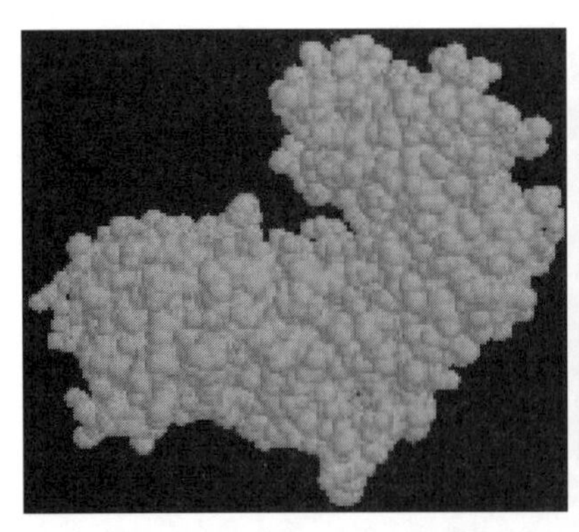

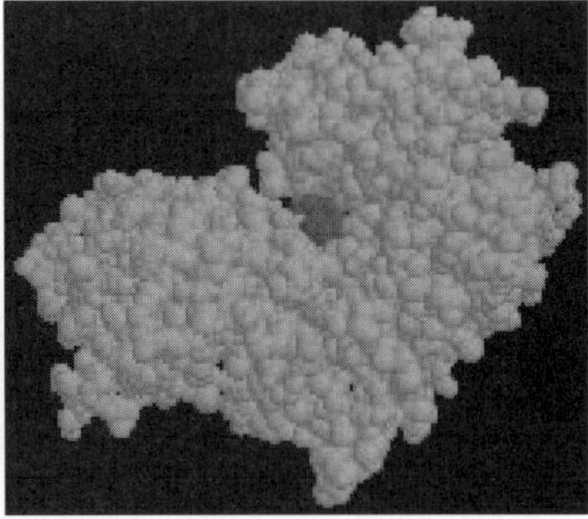

▲ **Figure 2.21** Notice how the active site changes shape as the glucose enters the active site of the enzyme hexokinase.

To convert substrate(s) into product(s), bonding within and between molecules must change. Breaking chemical bonds requires energy. Normally heating the substrates would provide the energy to start the reaction, known as the **activation energy**. (Think about when you want to start a bonfire – the reaction between the chemicals in wood and oxygen – you must first provide some energy to start the fire.) The addition of heat energy causes the agitation of atoms within the molecules; the molecules become unstable and the reaction can occur. In cells enzymes reduce the amount of energy needed to bring about a reaction; this allows reactions to occur without raising the temperature of the cell. The specific shape of the enzyme active site and complementary substrate(s) causes electrically charged groups on their surfaces to interact. The attraction of oppositely charged groups may distort the shape of the substrate(s) and assist in the breaking of bonds or formation of new bonds. In some cases, the active site may contain amino acids with acidic side chains; the acidic environment created within the active site may provide conditions favourable for the reaction.

2.3 What goes wrong in cystic fibrosis and other genetic diseases?

A mistake in translation can produce mRNA with one or more incorrect codons which may result in the production of a faulty protein or no protein at all. But because the fault is in the mRNA it would only affect the proteins produced from this one mRNA strand in this one cell, on this one occasion. It would not produce the problems seen in every epithelial cell of cystic fibrosis suffers. It is errors in the DNA that are responsible for inherited genetic conditions. These mistakes arise when DNA copies itself during the process of DNA replication.

DNA replication

When a cell divides, an exact copy of the DNA must be produced so that each of the daughter cells receives a copy. This process of copying the DNA is called **replication**. The DNA double helix unwinds from one end and the two strands split apart as the hydrogen bonds between bases break (Figure 2.22). Free DNA nucleotides line up alongside each DNA strand and hydrogen bonds form between the complementary bases. The enzyme DNA polymerase links the adjacent nucleotide to form a complementary strand. Thus, each strand of DNA acts as a **template** on which a new strand is built and, overall, two complete DNA molecules are formed. These are identical to each other and to the original DNA molecule. Check this by comparing them in Figure 2.22. Each of the two DNA molecules now contains one 'old'

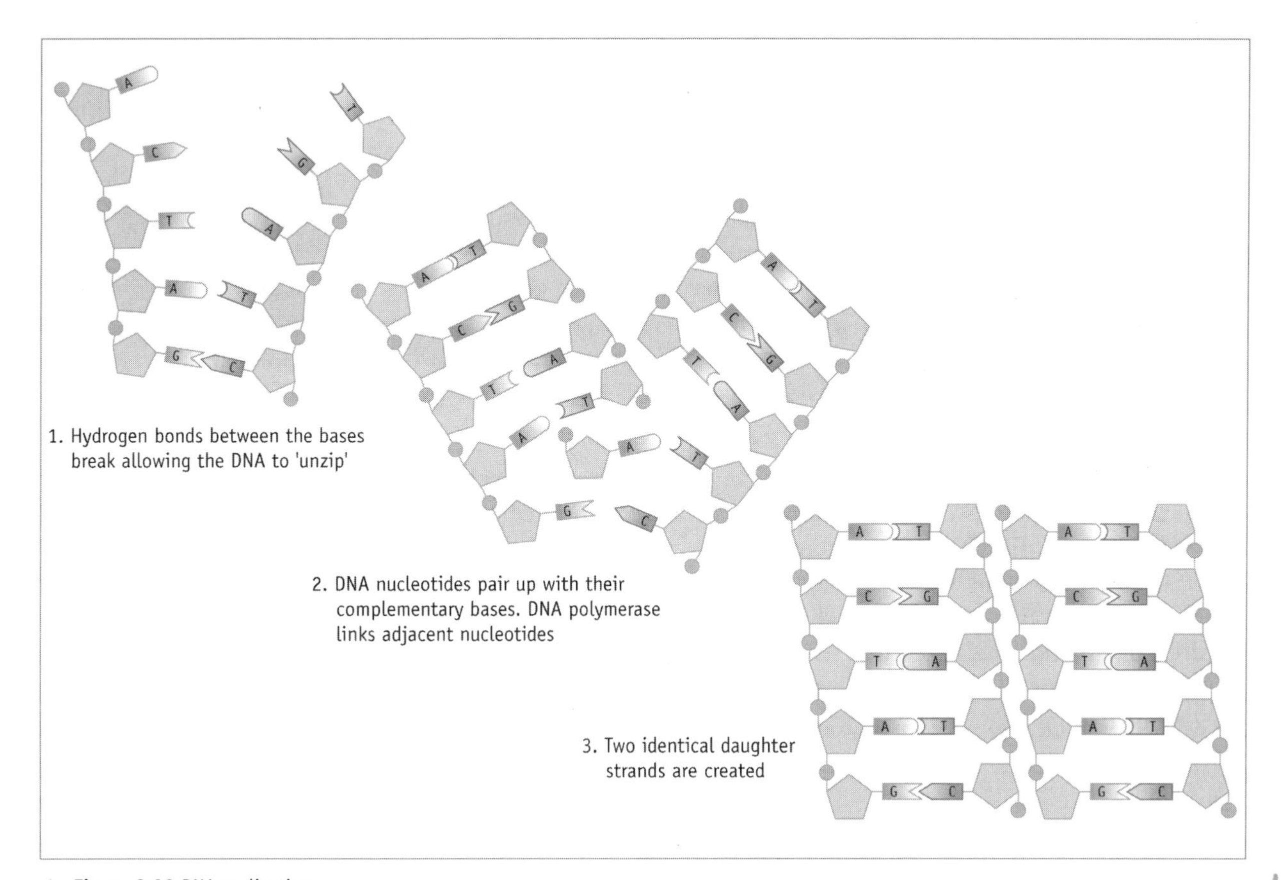

▲ **Figure 2.22** DNA replication.

strand and one 'new' strand. This process is therefore known as **semi-conservative replication**.

So what has all this to do with cystic fibrosis? Sometimes DNA replication does not work perfectly. As the 'new' strand of DNA is being built, an incorrect base may slip into place. This is an example of a **gene mutation**.

Q2.23 Assuming no mutations, what would be the sequence of bases on the complementary strand created by replication of an 'old' strand with the sequence C A G T C A G G C?

Q2.24 Identify the mutations that have occurred in replication of the same sequence in each of the following:
a) G T C A G G C C G
b) G A C A G T C C G
c) G T C A T G C C G
d) G T C G T C C G
e) G T C A G G T C C G

If a mutation occurs in the DNA of an ovary or testis cell that develops into an egg or sperm, it may be passed onto the next and subsequent generation. Some mutations may have no effect. Large amounts of the DNA found in the cell does not actually play a role in protein synthesis and therefore mutations that occur in these sections may have no effect. If a mutation occurs within a gene and a new triplet base created codes for a stop signal or a different amino acid, the protein formed can be faulty. This may cause a genetic disorder.

In the disease **sickle cell anaemia**, the DNA molecule that codes for one of the polypeptide chains in haemoglobin, the pigment in red blood cells which carries oxygen around the body, has a mutation. The base adenine replaces thymine at one position along the chain. The mRNA produced from this DNA contains the triplet code GUA rather than GAA. As a result the protein produced contains a glutamic acid rather than valine at this point. This small change has a devastating effect on the functioning of the molecule. The haemoglobin is less soluble and when oxygen levels are low the molecule forms long fibres that stick together inside the red blood cell distorting its shape. The half moon (sickle) shaped cells carry less oxygen and can block blood vessels.

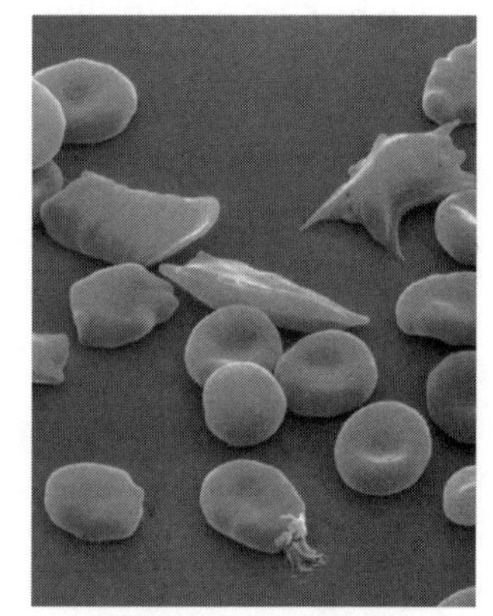

▲ **Figure 2.23** Can you identify those red cells that have been affected by sickle cell anaemia?

Cystic fibrosis is not as straightforward a genetic story as sickle cell anaemia. In the CF gene hundreds of different mutations have been identified that can give rise to cystic fibrosis. The mutations affect the CFTR protein in different ways. In some cases the ATP is unable to bind and open the ion channel; in other cases the channel is open but changes in the protein structure lead to reduced movement of chloride ions through the channel. The mutations are passed from parent to offspring. The most common mutation, known as the ΔF508 mutation, is the deletion of three nucleotides (Figure 2.24). This causes the loss of phenylalanine, the 508th amino acid in the CFTR protein.

Q2.25 Look at Figure 2.24 and explain why it is only phenylalanine that is lost even though two triplet codes have lost nucleotides.

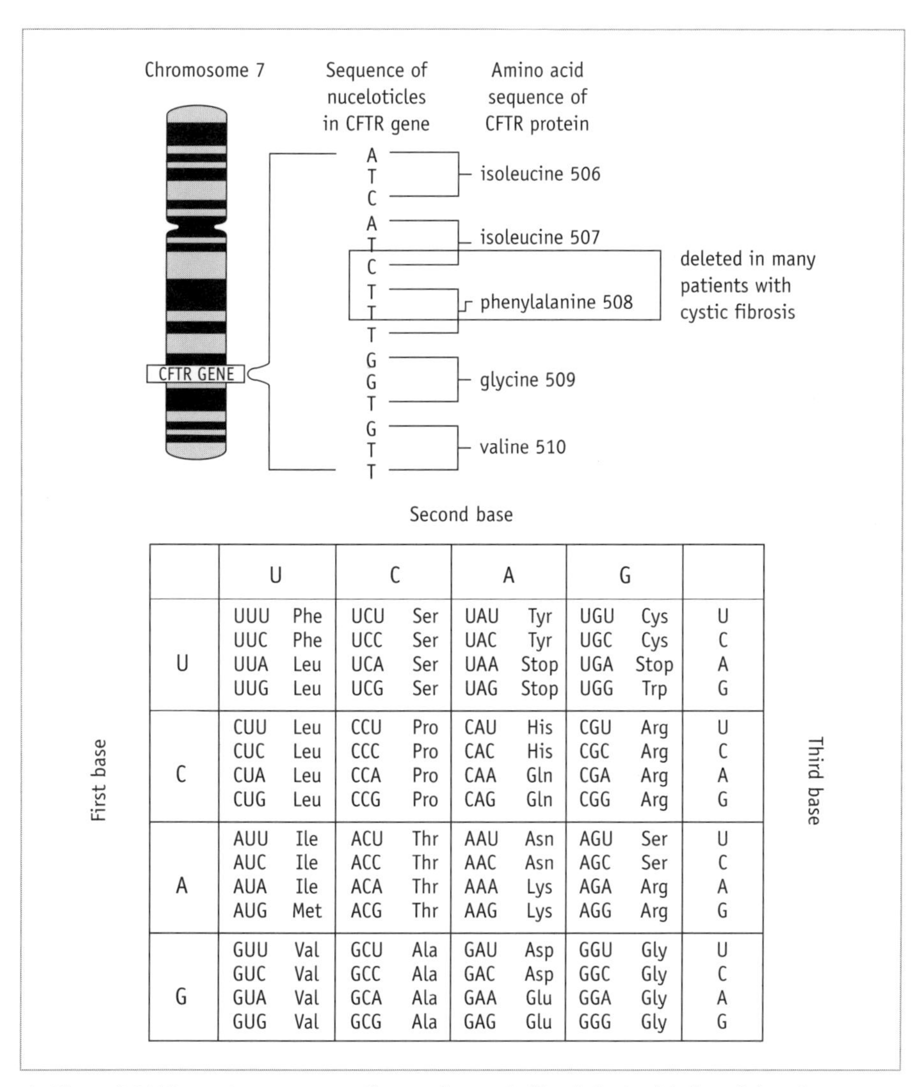

First base	U		C		A		G		Third base
U	UUU	Phe	UCU	Ser	UAU	Tyr	UGU	Cys	U
	UUC	Phe	UCC	Ser	UAC	Tyr	UGC	Cys	C
	UUA	Leu	UCA	Ser	UAA	Stop	UGA	Stop	A
	UUG	Leu	UCG	Ser	UAG	Stop	UGG	Trp	G
C	CUU	Leu	CCU	Pro	CAU	His	CGU	Arg	U
	CUC	Leu	CCC	Pro	CAC	His	CGC	Arg	C
	CUA	Leu	CCA	Pro	CAA	Gln	CGA	Arg	A
	CUG	Leu	CCG	Pro	CAG	Gln	CGG	Arg	G
A	AUU	Ile	ACU	Thr	AAU	Asn	AGU	Ser	U
	AUC	Ile	ACC	Thr	AAC	Asn	AGC	Ser	C
	AUA	Ile	ACA	Thr	AAA	Lys	AGA	Arg	A
	AUG	Met	ACG	Thr	AAG	Lys	AGG	Arg	G
G	GUU	Val	GCU	Ala	GAU	Asp	GGU	Gly	U
	GUC	Val	GCC	Ala	GAC	Asp	GGC	Gly	C
	GUA	Val	GCA	Ala	GAA	Glu	GGA	Gly	A
	GUG	Val	GCG	Ala	GAG	Glu	GGG	Gly	G

Key:

Ala = alanine
Arg = arginine
Asn = asparagine
Asp = aspartic acid
Cys = cysteine
Gln = glutamine
Glu = glutamic acid
Gly = glycine
His = histidine
Ile = isoleucine
Leu = leucine
Lys = lysine
Met = methionine
Phe = phenylalanine
Pro = proline
Ser = serine
Thr = threonine
Try = tryptophan
Tyr = tyrosine
Val = valine

▲ **Figure 2.24** The most common mutation causing cystic fibrosis is the deletion of three bases within the CF gene. You can see the consequences of this by consulting the genetic dictionary beneath.

Do Claire and Nathan have one of these mutations? Will they pass it on and if they do, will the child have the disease?

You can find out more about the mutations that cause cystic fibrosis by visiting the Oxford Gene Medicine Research Group web site.

2.4 How is cystic fibrosis inherited?

Claire and Nathan know that cystic fibrosis is inherited but they need to understand *how* it is inherited. This will let them know the chances of their children inheriting the disease.

A gene is a length of DNA that codes for a protein. Every cell (except the sex cells) contains two copies of each gene. For any particular gene the two copies are located in the same position, or **locus**, on one pair of **homologous chromosomes**, one of the 23 pairs of chromosomes found in our cells. Within each of our 23 pairs of homologous chromosomes, one chromosome comes from our mother and the other from our father. Cystic fibrosis is cause by a gene mutation passed on from parents to their children. But consider the following three situations:

1 A couple have six children. The first five are healthy, the sixth has cystic fibrosis. Neither parent has the disease.

2 Another couple have two children, both of whom have cystic fibrosis. Again, neither parent has the disease.

3 A woman with cystic fibrosis is told that it is very unlikely that her children will have the disease, but they will all be 'carriers' and could have children with the illness.

So what is going on?

As we have seen, cystic fibrosis is caused by a mutation in the length of DNA that codes for the CFTR protein, the CF gene. The CF gene occurs in two alternative forms or **alleles**. Firstly, there is the normal allele which causes the production of functioning CFTR protein; this can be represented by the letter F. Secondly, there is the mutated allele which produces a non-functional protein; this can be represented by the letter f. Since every human being contains two copies of this gene in each cell there are three possible combinations that can occur, namely:

1 FF – This person has two identical copies of the normal allele and does not have cystic fibrosis

2 ff – This person has two copies of the mutated allele and does have cystic fibrosis

3 Ff – This person has one normal allele and one mutated allele. He or she does not have cystic fibrosis but is a **carrier**. He or she could have children who have the disease.

Genotype	Phenotype
FF	Normal
ff	Cystic fibrosis
Ff	Normal, but carrier

▲ **Table 2.1** The relationship between genotype and phenotype at the cystic fibrosis gene

The alleles that a person has are known as their genotype. Persons 1 and 2 have a **homozygous** genotype for the CF gene, as they have two identical copies of an allele. Person 3 has two different alleles and is **heterozygous** for the CF gene.

The characteristic caused by the genotype, i.e. its observable effect, is the **phenotype**. Table 2.1 summarises the cystic fibrosis situation.

F is called the **dominant** allele. It affects the phenotypes of both the homozygote (person who has the homozygous genotype, FF) and the heterozygote (person with the heterozygous genotype, Ff). On the other hand, f is **recessive**; it only affects the phenotype of the homozygote (ff).

If two CF carriers had children what genotypes and phenotypes would we expect their children to have?

In the UK about one person in twenty four is a cystic fibrosis carrier. They can pass the disease onto their children because when gametes are produced each egg or sperm will contain only *one* allele, in this case either F or f. These two types of gametes are produced in roughly equal numbers. The expected genotypes of children produced by two cystic fibrosis carriers can be shown in a genetic diagram, known as a Punnett square (Table 2.2). This illustrates all the possible ways in which the two types of allele can combine, and thus shows the possible genotypes that can occur in the children.

Parents' phenotypes	Normal	Normal
Parents' genotypes	**Ff**	**Ff**
Gamete genotypes	(F) or (f)	(F) or (f)

		Gametes from mother	
		F	f
Gametes from father	F	FF	Ff
	f	Ff	ff

▲ **Table 2.2** Punnett squares can be used to work out offspring genotypes from parental genotypes

It is important to understand that this means that every time a child is born to parents both of whom are carriers of cystic fibrosis there is a 1 in 4 or 0.25 probability that the baby will have the genotype ff and suffer from cystic fibrosis. There is also a 1 in 4 probability that the genotype will be FF, and a 1 in 2 (2 in 4) or 0.5 probability that it will be Ff, a CF carrier.

Until very recently people with the genotype ff did not often survive to be adults and thus did not have any children. You might think that natural selection would have eliminated such a harmful gene. In fact, heterozygotes (Ff) have some protection against the dangerous disease typhoid. So in areas where this illness was common they would have been at a definite advantage.

Q2.26 Give a genetic explanation for the three situations described at the top of page 82.

Cystic fibrosis is an example of **monohybrid inheritance**, so called because the characteristic is controlled by only one gene. Most human characteristics are inherited in a much more complex way, and are often influenced by environmental factors. However, there are a few characteristics controlled by single genes.

For example, **thalassaemia** is a genetic disease caused by recessive alleles of a gene on chromosome 11. The gene is involved in the manufacture of the protein haemoglobin, found in red blood cells, which carries oxygen around the body. There are a number of different mutations which can affect this gene. Someone who is homozygous for one of these conditions either makes no haemoglobin at all, or makes haemoglobin that cannot carry out its function. The homozygous condition is often eventually lethal. People who are heterozygous show no symptoms, but have some protection against malaria. Thus, this condition is relatively common in people who live (or have ancestors who lived) in areas where malaria occurs or occurred in historical time, particularly around the Mediterranean Sea. Because of this 'heterozygous advantage' the mutant alleles have not disappeared.

Activity

You might use Reebops to investigate inheritance in **Activity 2.15. ASO2ACT15**

Other conditions such as **albinism**, **phenylketonuria** and **sickle cell anaemia** are also caused by single recessive alleles. **Achondroplasia**, on the other hand, is caused by a dominant allele. A homozygote, carrying two copies of the allele for achondroplasia, always dies. Someone who is heterozygous for this condition will show very restricted growth, usually attaining a height of about 125–130 cm. The trunk is of average length but the limbs are much shorter. There is absolutely nothing wrong with their

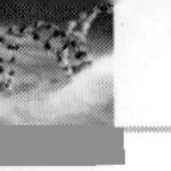

mental powers. **Huntington's disease** and the ability to taste PTC are also caused by dominant alleles.

You should not suppose that characteristics determined by a single gene are found only in humans. They occur in all organisms. Indeed, they were first discovered in the garden pea by Gregor Mendel in the 1850s and 60s.

Mendel is rightly known as the father of genetics. He was a monk and carried out a huge number of breeding experiments in the garden of his Moravian monastery. Mendel died without other scientists appreciating the significance of his work. However, he established that a number of characteristics of the garden pea were determined by separate genes. For example, whether the plants are tall (1.9–2.2 m in height) or short (0.3–0.5 m) is determined by one gene. Whether the seeds are smooth or wrinkled is determined by another gene.

Activity

Activity 2.16 lets you apply ideas about inheritance to some other situations.
AS02ACT16

2.5 Treatment of cystic fibrosis

The symptoms and severity of cystic fibrosis are, to a considerable extent, specific to each individual; this is true even if the individuals share the same mutation. The consequence of this is that the treatment for each individual needs to be specific to that individual. And it is difficult to predict how badly affected a child will be simply by looking at the symptoms of any CF relatives.

The CF treatments available alleviate some of the symptoms. This can contribute to an enhanced life span and improved quality of life but as yet there is no known cure for cystic fibrosis. There is a range of treatment options available for a person with cystic fibrosis.

1 Physiotherapy

Rhythmical tapping of the walls of the chest cavity (percussion therapy) can help loosen the mucus and improve the flow of air into and out of the lungs (Figure 2.25). Such treatment needs to be carried out regularly, twice a day.

▲ **Figure 2.25** Children learn how to perform their own physiotherapy. How can hitting the back and front of the chest cavity help?

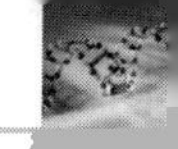

2 'Flutter' device

The 'flutter' is a small, hand-held device that looks like a pipe and which the patient can use without assistance. When the person exhales through the flutter, a special valve causes rapid air pressure fluctuations in their trachea, bronchus and bronchioles. The resulting vibrations dislodge the mucus and aid mucus movement.

3 Digestive enzyme supplements

If the pancreatic duct is blocked, the food molecules in the small intestine cannot be broken down far enough for full absorption. The enzyme supplements help to complete the process of digestion.

Q2.27 What type of enzymes would you expect to find in such supplements?

4 Medication

There are a wide range of medications commonly used to relieve the symptoms of CF. These include:

- Bronchodilators
  These drugs are inhaled using a nebuliser. They relax the muscles in the airways, opening them up and relieving tightness of the chest.
- Antibiotics
  The build up of mucus in the lungs can lead to regular infections, especially by bacteria belonging to the genus *Pseudomonas*, requiring antibiotic treatment. Many different antibiotics are often needed because of the range of pathogens found in the lungs. Long-term treatment with antibiotics can have side effects, for example nausea and vomiting. Regular use of antibiotics can also lead to a build up of resistance which makes the treatment less effective.
- DNAase enzymes
  Infection of the lungs leads to the accumulation of white blood cells in the mucus. The breakdown of these white blood cells releases DNA which can add to the 'stickiness' of the mucus. The DNAase enzymes inhaled using a nebuliser break down the DNA, so the mucus is easier to clear from the lungs.
- Steriods
  Steroids are used to reduce inflammation of the lungs.

5 Diet

Careful choice of what to eat can reduce the individual's dependence on the use of enzyme supplements, as well as improving their growth and development. It can also increase their strength and resistance to disease. Extra energy and protein should be eaten, to overcome the reduced uptake. Adults with CF are recommended to eat high fat and carbohydrate foods and their diet should include double the quantity of protein recommended for people who do not have CF.

6 Heart and lung transplant

If the lungs become very damaged and their efficiency is dramatically reduced, other treatments may become ineffective at relieving the symptoms. The only option available may be to replace the damaged lungs with a heart and lung transplant.

Q2.28 Suggest why doctors would give a heart and lung transplant rather than just a lung transplant.

Activity

In **Activity 2.17** you can try recommending treatment options for individuals with CF. **ASO2ACT17**

Future prospects of CF treatment

These treatments only reduce some of the effects of the disease. Understanding the nature of the gene involved in cystic fibrosis has raise the possibility of a cure in the future through **gene therapy**. Effective gene therapy would treat the cause rather than the symptoms of the disease, but what is gene therapy?

In gene therapy the genotype of target cells (those affected by the disease) is altered by adding the 'normal' allele of the gene. Trials of CF gene therapy started in the UK back in 1993 and they have successfully transferred the 'normal' CFTR allele to the lung epithelial cells of CF patients. A plasmid–liposome complex is used to transfer the gene into the lungs as follows (Figure 2.26).

First, a copy of the normal allele is inserted into a loop of DNA (called a **plasmid**). The plasmids are then combined with **liposomes** (spherical phospholipid bilayers). The positively charged head groups of the phospholipids combine with the DNA (a weak acid and so negatively charged) to form the liposome-DNA complex. The CF patient breathes in an aerosol containing these complexes using a nebuliser. The liposomes fuse with the cell membrane and carry the DNA into the cell. The normal CF allele is transferred to the nucleus where it is transcribed. A functioning CFTR protein is produced and incorporated into the cell membrane, thus restoring the ion channel and avoiding the symptoms of CF.

In trials the successful transfer and expression of the gene has been achieved. The presence of functioning ion channels is indicated by a reduction in the potential difference across the membranes in the nose and lungs, and increased chloride secretions. The results of trials have so far demonstrated a correction of the chloride secretion but not the associated sodium secretion problems. In trial results published in 1999, chloride transport in the lungs was restored to 25% of normal. However, this type of improvement is temporary. Cells are continuously lost from the epithelium lining the airway so the transfer of the allele to these cells does not offer a permanent solution. The longest the correction has lasted in any trial is about 15 days so this treatment would have to be repeated throughout the

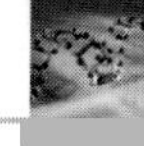

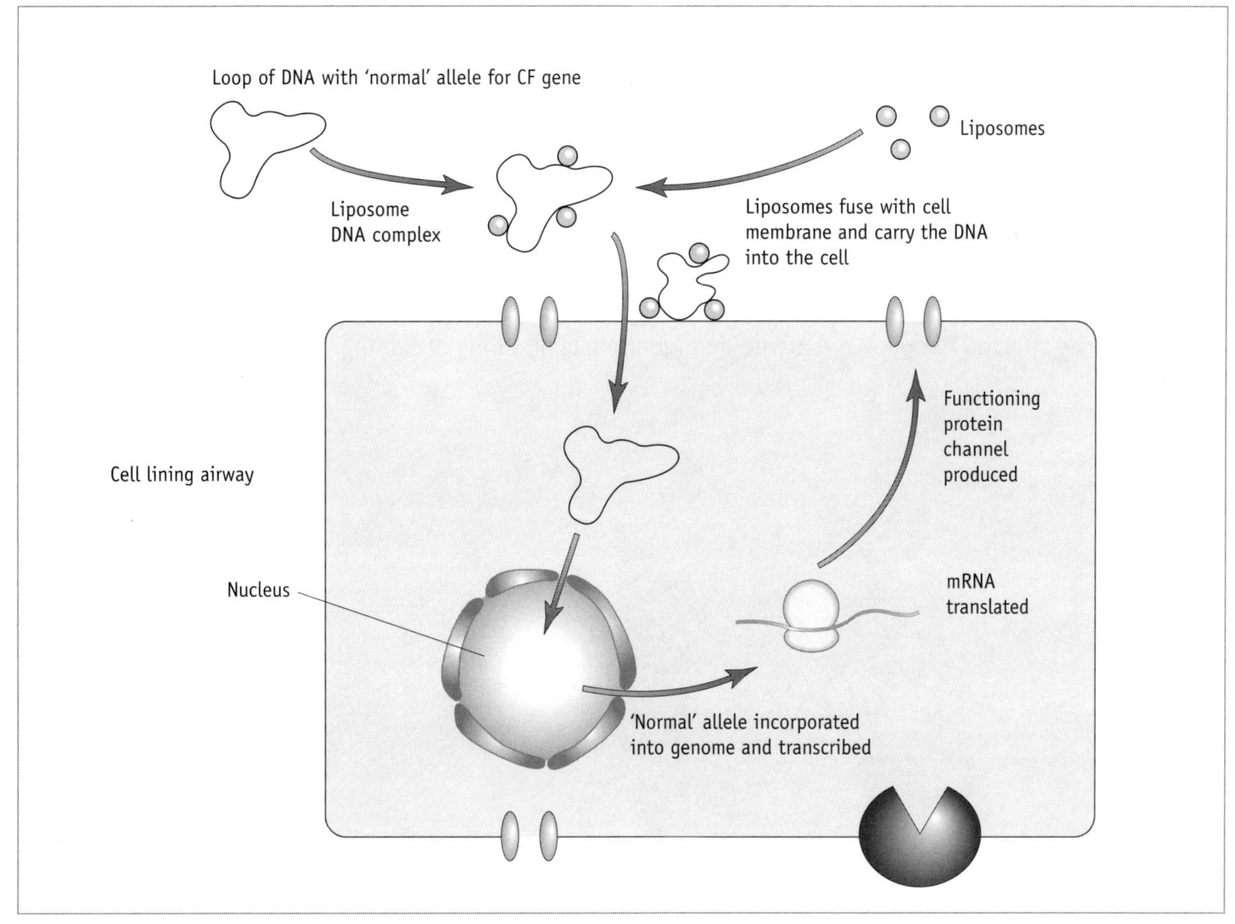

▲ **Figure 2.26** Gene therapy for cystic fibrosis may yet be successful but difficulties with the delivery of the gene mean that the trials must continue.

patient's life. Trials continue with the aim of improving the delivery of the gene and thus increasing the effectiveness of the treatment that may one day offer a cure.

The only condition in humans to have been successfully treated using gene therapy by 2002 was a very rare disorder called severe combine immune deficiency (SCID). Pateints with this disease cannot make a particular enzyme needed for the immune system to work. In 1990, white blood cells were removed from a girl with the disease. The alleles for the functioning gene were inserted into her white cells using a virus. The cells were then replaced. The girl has to have regular transfusions with the modified white cells but otherwise is fit and well. It is hoped that gene therapy will help in the treatment of thalassaemia, haemophilia and sickle cell disease but to date only limited progress has occurred with these diseases.

These treatments are all concerned with altering specific **somatic cells** (body cells) and are permitted under UK legislation. The alternative approach of altering the **germ cells** (sperm or eggs) so that every cell in the body contains the new gene is not permitted. There are ethical objections because of concerns about possible effects in future generations when the new gene is inherited.

2.6 Testing for CF

Even though one person in 25 in the UK are CF carriers, the first time that most people realise that they carry the CF allele is when one of their children is born with the disease.

Activity

Activity 2.18 looks at sweat analysis of people with CF. **AS02ACT18**

Once cystic fibrosis is suspected, it is possible to carry out conventional tests to confirm the diagnosis. Ancient European folklore warned that a child tasting salty when kissed would die young, and this has become the basis of a modern test that measures the level of salt in sweat (Figure 2.27). The test works because affected people have markedly higher concentrations of salt in their sweat.

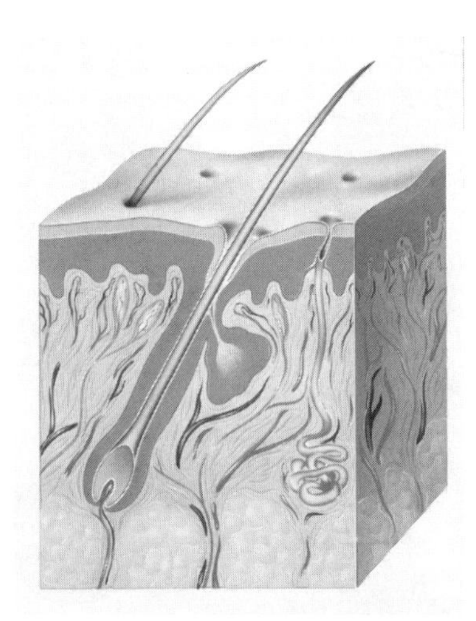

▲ **Figure 2.27** In normal sweat glands salt is reabsorbed from sweat before it reaches the skin, leaving it only slightly salty. With an abnormal CFTR protein, much more salt remains in the sweat.

A drawback to this form of testing is that the actual levels of salt vary in all individuals, so it is possible to incorrectly diagnose CF in normal individuals ('**false positives**') and also to consider some mild forms of CF normal ('**false negatives**').

Q2.29 What could be the consequences when false positives and false negatives occur?

A blood test (immunoreactive trypsinogen test (IRT)) can also be used. This is used on small babies who do not produce enough sweat to test and also to confirm positive sweat tests. The test detects the presence of the protein trypsinogen in the blood. In some countries, such as Australia and parts of the USA, all newborn babies are routinely tested, since early diagnosis allows treatment to begin immediately, which may improve their health in later years. Testing of all new-born babies was introduced in Scotland in April 2002 and there are plans to have all UK babies routinely tested by 2004.

Although there may be some benefits of this type of testing, it cannot prevent the disease, and it can only be used on people who are already affected. The sequencing of the gene responsible for the CFTR protein in 1989 led to the possibility of **genetic testing**, where the abnormal gene can be identified in the DNA of any cells. This in turn has paved the way for

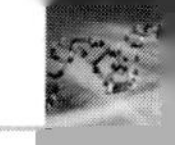

genetic screening to identify carriers, and also to diagnose CF in an embryo or fetus. (The terms 'genetic testing' and 'genetic screening' are sometimes distinguished but often, as here, used interchangeably.)

How is genetic testing done?

Analysing the DNA

Genetic testing can be performed on any DNA, so it is possible to take samples of cheek cells, white blood cells or cells obtained from a fetus or embryo. Only known base sequences can be tested, so the test is performed for the most common mutations that cause cystic fibrosis.

Q2.30 What does the fact that the test is only performed for the most common mutations that cause CF mean about the reliability of the test? Will there be any false positives or negatives?

Once obtained, the DNA sample is tested by the method illustrated in Figure 2.28 and described below. The essence of this approach applies to many genetic tests.

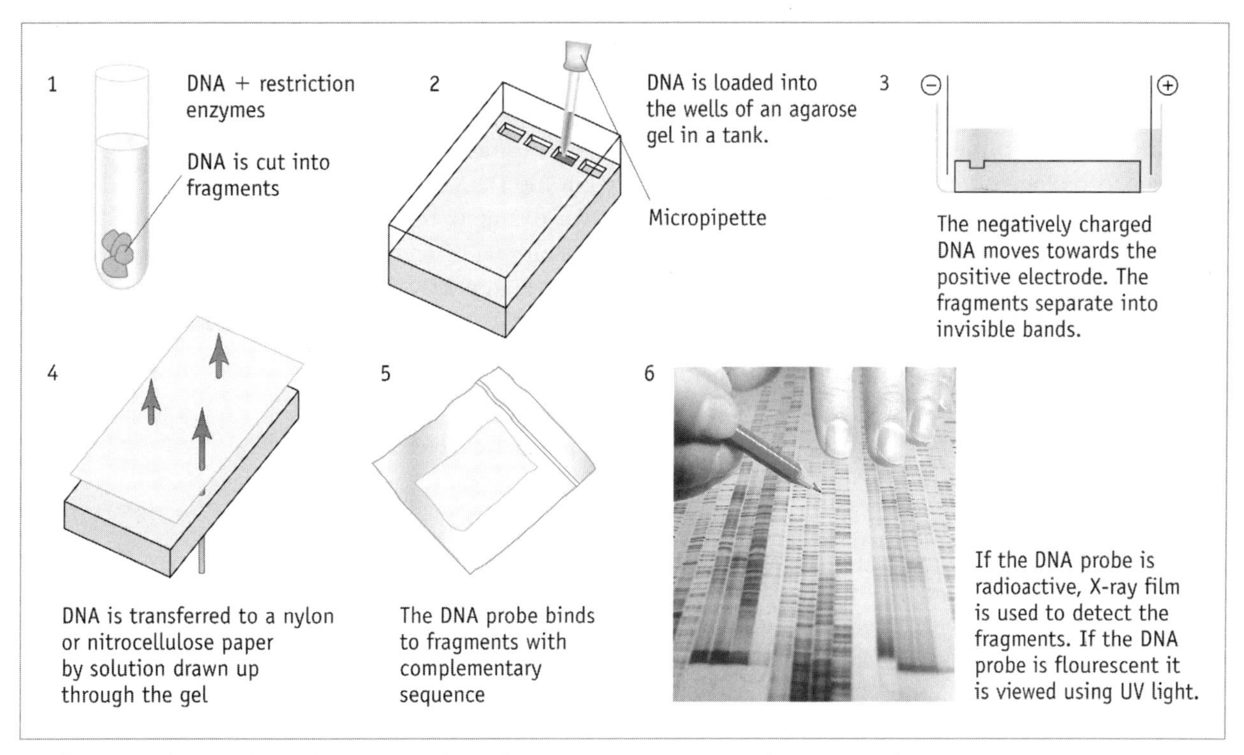

▲ **Figure 2.28** The steps in carrying out a genetic test for the CF mutation or many other gene mutations.

1 **DNA extraction** removes DNA from the cells.
2 **Restriction enzymes** are used to cut the DNA into fragments.
3 **Gel electrophoresis** is used to separate the fragments according to their size.
4 **Southern blotting** is used to transfer the DNA fragments to a nylon or nitrocellulose filter. The filter is placed over the gel and then dry adsorbent paper is used to draw the DNA fragments from the electrophoresis gel onto the nylon or nitrocellulose. The paper acts as a

wick to draw an alkaline buffer solution through the gel. The DNA strands separate, exposing the base sequences.

5 **Gene probes** are short sequences of DNA that have a complementary base sequence to the gene being sought. In the case of cystic fibrosis this will be complementary to a known mutant sequence. Traditionally, the probe is made using radioactive phosphorus (^{32}P), making the DNA fragment radioactive. More recent methods use DNA probes that have had fluorescent molecules chemically bound to them. Large quantities of the probe are added to the filter and allowed time to bind with complementary sequences (hybridisation) before any unbound probe is washed away.

6 **Determining if the probe has bound to any DNA fragments**. In the case of radioactive probes, the nylon or nitrocellulose filter is dried and placed next to X-ray film. Any radioactive probe present will expose or blacken the film, showing that the DNA sequence being looked for is present. For fluorescent probes, the filter can be viewed under ultra-violet light to reveal the bound DNA.

Restriction enzymes

The genetic screening test uses a type of enzyme called a **restriction enzyme**, more correctly called a restriction endonuclease, to cut DNA into fragments. Restriction enzymes are found naturally in bacteria where their function is to cut up invading viral DNA. The value of these restriction enzymes to biologists lies in the fact that they will only cut the DNA at specific base sequences (usually just a few bases long). For example, the enzyme EcoRI comes from the bacterium *Escherichia coli* strain RY13 and it splits the DNA wherever the sequences of bases is GAATTC. Figure 2.29 shows where it cuts the DNA.

▲ **Figure 2.29** EcoRI always cuts the DNA at this site. Each restriction enzyme cuts at a specific sequence.

In nature, the bacteria are able to protect their own DNA from the enzymes by modifying the bases in the specific sequence targetted by their own restriction enzymes. Restriction enzymes are also used in preparing DNA for gene therapy.

Gel electrophoresis

The fragments produced by restriction enzymes are separated by **gel electrophoresis**. The DNA fragments are placed on a gel made of agarose or polyacrylamide, both of which provide a stable medium through which the DNA molecules can move. The gel is connected to electrodes that produce an electrical field, and the DNA fragments migrate in the field according to their overall charge and size. Smaller fragments travel faster and therefore further in a given time.

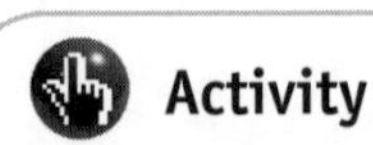

Activity

Try out these techniques using the simulation in **Activity 2.19, AS02ACT19** or practically in **Activity 2.20, AS02ACT20**

How can genetic screening be used?

It might seem that the genetic testing could be used to make certain that a diagnosis of CF is correct. However, since there are a large number of different mutations of the CFTR gene which cause the disease, a negative result must be treated with caution. It is not currently feasible to test for all of the hundreds of possible mutations.

A particular value of genetic testing is to identify the presence of the cystic fibrosis allele where no symptoms are apparent, such as in a carrier, or in an embryo or fetus. In this way it may be possible to avoid having a child with cystic fibrosis, or to be prepared well in advance for any problems.

Activity

You might use the techniques in the genetic screening simulation Nature's Dice **Activity 2.21, ASO2ACT21.**

Identification of carriers

A sample of blood or cells taken from inside the mouth can be used to detect abnormal alleles in people without the disease who are heterozygous. Where there has been a history of cystic fibrosis in a family this can be of value in assessing the probability of having a child with the disease. **Counselling** is offered before and after testing, and parents can make informed decisions about how to proceed.

Because the testing has become a routine procedure, there have been moves to make the testing for carriers more widespread, and some people advocate screening of all potential parents. There are a number of issues to consider, however, including cost and whether sufficient counselling is available.

Prenatal testing

Looking at the genetic make up of an embryo or fetus is now a common part of care during pregnancy. For example, checking for chromosomal disorders such as Down's syndrome has been carried out for many years. With the development of genetic testing for individual genes the possibilities are now more extensive, and this includes checking the genotype of a child with respect to cystic fibrosis before it is born.

Currently, there are two techniques used for obtaining a sample of cells from the child. The more common is **amniocentesis**, which involves inserting a needle into the amniotic fluid surrounding the fetus to collect cells that have fallen off the placenta and fetus (Figure 2.30). This can be carried out at around 15–17 weeks of pregnancy, and involves a risk of between 0.5% and 1% of causing a miscarriage. **Chorionic villus sampling (CVS)** removes a small sample of placental tissue, which includes cells of the fetus, either through the wall of the abdomen or through the vagina. This can be carried out earlier, between 8 and 12 weeks, since there is no need to wait for amniotic fluid to develop, but it carries a risk of about 1 to 2% of inducing a miscarriage.

It is likely that in the future it will be possible to collect fetal cells from the blood of the mother with no risk to the fetus. When this happens, genetic screening on a large scale could become even easier.

Whether amniocentesis or CVS is used, where the genetic test proves positive for the disease, one possibility is for the woman to have an abortion. Having an abortion is easier for the woman, both physically and emotionally, in the earlier weeks of pregnancy. This can mean that the higher risk of miscarriage associated with CVS are worthwhile.

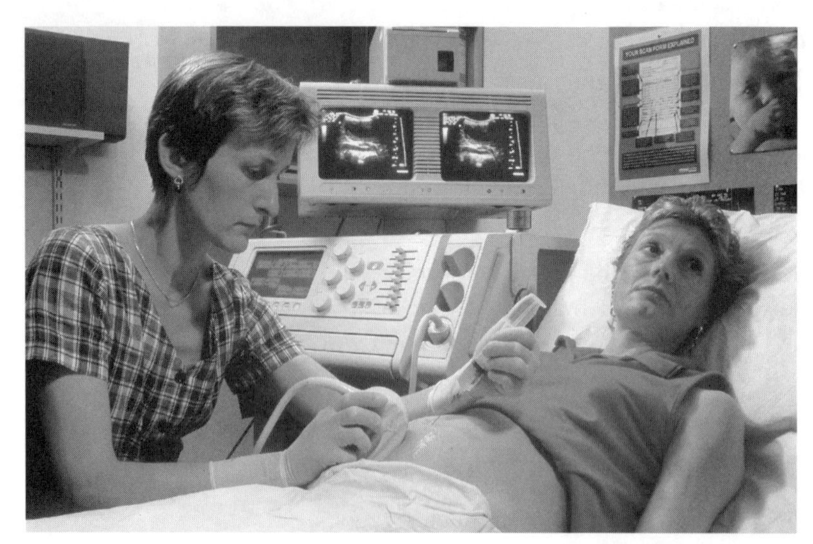

▲ **Figure 2.30** To complete a prenatal test during pregnancy involves techniques that do carry some risk.

Q2.31 Are there circumstances where people might choose not to test, even if the test was offered to them?

It is also possible to test an embryo before it has implanted in the uterus when ***in vitro* fertilisation (IVF)** is carried out. It is possible to remove a cell from an embryo growing in culture when it has only 8 or 16 cells without harming the embryo at all (Figure 2.31). The DNA of the cell can be analysed and the results used to decide whether to place the embryo into the womb. IVF, however, is still an expensive and fairly unreliable procedure, and although this avoids the need for abortion, this approach is not used routinely.

Activity

Activity 2.22 looks at issues involved in using widespread carrier screening. **AS02ACT22**

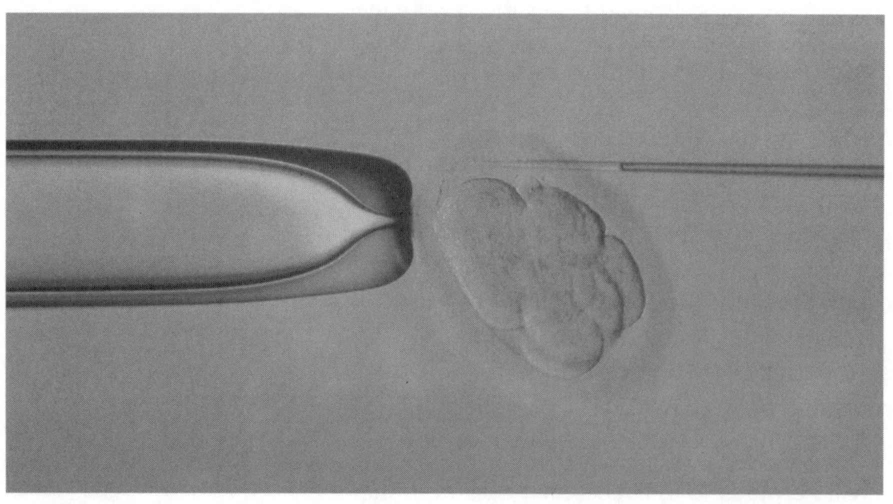

▲ **Figure 2.31** Why can one cell be removed at this stage without harming the developing embryo? We will see the answer to this question in Topic 3.

What is right and what is wrong?

How should we decide in life what is right and what is wrong? For example, should we always tell the truth? Can we ever justify turning down a request for help? Should Claire and Nathan have an abortion if they find that their unborn baby has cystic fibrosis?

All of us have **moral** views about these and other matters. For example, you might hold that lying and abortion are always wrong and helping people always right. But in order to maintain that something is **ethically** acceptable or unacceptable, you must be able to provide a reasonable explanation as to *why* that is the case.

There is no one universally accepted way of deciding whether something is ethically acceptable or not. What there are instead are a number of **ethical frameworks** each of which allows you to work out whether a particular action would be right or wrong if you accept the ethical principles on which the framework is based. Usually you get the same answer whichever framework you adopt. But not always! This is why perfectly thoughtful, kind and intelligent people sometimes still disagree completely about whether a particular course of action is justified or not.

We will examine four widely used ethical frameworks. You should find these of value when considering various issues, such as genetic screening and abortion, raised in this topic. We will also refer to these frameworks in other topics in this course.

1 Rights and duties

Most of us tend to feel that there are certain human **rights** that should always be permitted. For example, we talk about the right to life, the right to a fair trial and the right to freedom of speech. Certain countries, for example the USA, have some of these rights enshrined in their constitutions.

If you have a right to something, then I may have particular **duties** towards you. For example, suppose that you are a six month old baby with a right to life and I am your parent. I have a duty to feed you, wash you, keep you warm and so on. If I don't fulfil these duties, I am failing to carry out my responsibilities and the police or social services may intervene.

But where do rights come from? Some people with a religious faith find them in the teachings of their religion. For example the ten commandments in the Jewish scriptures talk about not stealing, not murdering, telling the truth and so on.

But nowadays, of course, many people, indeed in the UK most people, have little or no religious faith. So where can they – perhaps you – find rights? The simplest answer is that rights are social conventions built up over thousands of years. If you want to live in a society you have, more or less, to abide by its conventions.

2 Maximising the amount of good in the world

Perhaps the simplest ethical framework says that each of us should do whatever **maximises the amount of good in the world**. For example, should I tell the truth? Usually yes, as telling lies often ends up making people unhappy and unhappiness is not a good. But sometimes telling the truth can lead to more unhappiness. If your friend asks you if you like the present they have just given you and you don't, would you tell the truth? Most of us would tell a 'white lie', not wanting to harm their feelings.

This ethical approach is known as **utilitarianism**. Notice that utilitarians have no moral absolutes. A utilitarian would hesitate to state that anything is always right or always wrong. There might be circumstances in which something normally right (e.g. keeping a promise) would be wrong and there might be circumstances when something normally wrong (e.g. killing someone) would be right.

3 Making decisions for yourself

One of the key things about being a human is that we can make our own decisions. There was, for example, a time when doctors simply told their patients what was best for them. Now, though, there has been a strong move towards enabling patients to act **autonomously**. People act autonomously when they make up their own mind about something. If you have ever had an operation, you will probably have signed a consent form. The thinking behind this is that it isn't good for a surgeon to be allowed to operate on you unless you have given **informed consent**.

Of course, it is perfectly possible autonomously to decide to be absolutely selfish! Autonomy clearly isn't the only good in the world. A utilitarian would say we need to weigh the benefits of someone acting autonomously with any costs of them doing so. Only if the overall benefits are greater than the overall costs is autonomy desirable. Someone who believes in rights and duties might say that each of us has a right to act autonomously but also has a duty to take account of the effects of our actions on others.

4 Leading a virtuous life

A final approach is one of the oldest. This holds that the good life (in every sense of the term) consists of acting **virtuously**. This may sound rather old-fashioned but consider the virtues that you might wish a good teacher/lecturer to have. She or he might be understanding, able to get you to learn what you want/need to learn and believe in treating students fairly.

Traditionally the seven virtues were said to be **justice**, **prudence** (i.e. wisdom), **temperance** (i.e. acting in moderation), **fortitude** (i.e. courage), **faith**, **hope** and **charity**. Precisely what leading a virtuous life means can vary. Think about the virtues you might like to see in a parent, a doctor and a girl- / boyfriend. What would be the virtuous course of action for Claire and Nathan?

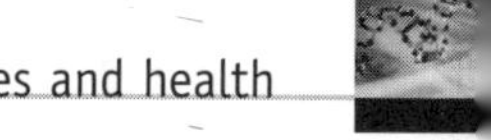

Genetic counselling

If a couple like Nathan and Claire are at risk of having a child with a genetic disorder, a genetic counsellor can provide advice. Such a person will help the couple understand how the disease is inherited and what the risk of any child they conceive has of having the disease. They will explain the tests available and the possible courses of action available depending on the outcome of the tests. The counselling should help the couple decide whether to be tested and whether to have children, use *in vitro* fertilisation with screening or use prenatal screening.

Q2.32 If both parents are found to be carriers what are the options open to them?

Activity

The role play in **Activity 2.23** lets you think about some of the issues covered in this topic. **AS02ACT23**

Summary

Having completed Topic 2 you should be able to:

- Explain the role of concentration gradients in the movement of molecules (including O_2, CO_2 and solutes) by diffusion.
- Describe the properties of gas exchange surfaces (large surface to volume ratio, thickness of surface, difference in concentration) and explain how the structure of the lung provides a large surface area to volume ratio.
- Describe the structure of the unit membrane (fluid mosaic model).
- Describe how the effect of temperature on membrane structure can be investigated practically.
- Explain what is meant by osmosis, active transport, passive transport and facilitated diffusion and the involvement of carrier and channel proteins in membrane transport.
- Describe the structures of DNA and messenger RNA, understand the nature of the genetic code and understand that a gene is a sequence of bases on a DNA molecule.
- Explain the process of protein synthesis (transcription, translation, tRNA, mRNA, ribosomes, the role of start and stop codons).
- Explain the significance of a protein's primary structure in determining its three-dimensional structure and properties.
- Explain the mechanism of action and specificity of enzymes in terms of their three-dimensional structure.
- Describe how enzyme concentrations can affect the rates of reactions and how this can be investigated practically.

- Describe the process of DNA replication.
- Explain how errors in DNA replication can give rise to mutations and explain how CF results from one of a number of possible gene mutations.
- Recall the terms genotype and phenotype and explain what is meant by monohybrid inheritance.
- Use a knowledge of genetics to answer questions about monohybrid inheritance including CF, albinism, thalassaemia and garden pea height and seed morphology.
- Explain how the expression of the CF gene impairs the functioning of the gaseous exchange, digestive and reproductive systems.
- Describe ways in which CF is currently treated (physiotherapy, drugs and gene therapy).
- Describe the principles of gene therapy and distinguish between somatic and germ line therapy.
- Discuss the moral and ethical issues involved in gene therapy.
- Describe how gel electrophoresis can be used to separate DNA fragments of different length.
- Describe how the genetic profiles produced by gel electrophoresis can be used in genetic screening using gene probes.
- Explain the uses of genetic screening in the identification of carriers, prenatal testing (amniocentesis and chorionic villus sampling) and embryo testing.
- Discuss the social, ethical, moral and cultural issues related to genetic screening.

Now that you have finished Topic 2, complete the end-of-topic test before starting Topic 3.
AS02RVT02

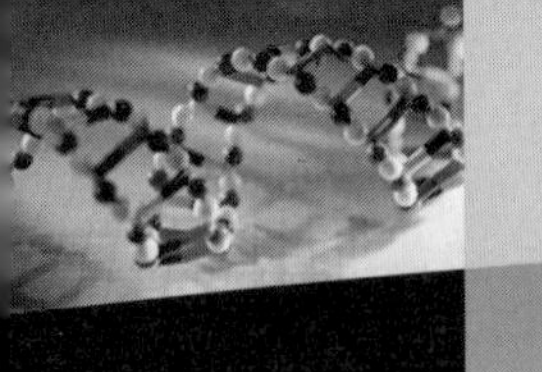

Answers

Answers to in-text questions for TOPIC 1

Q1.1	probability	per 1 000 000 people
Heart disease	0.23%	2300
Lung cancer	0.06%	600
Road accidents	0.006%	60
Accidental poisoning	0.0017%	17
Injury purposely inflicted by another	0.00049%	4.9
Railway accidents	0.00007%	0.7
Lightning	nearly 0	0.1

Q1.2 Movement of oxygen; carbon dioxide; and other products carried by blood; relies on diffusion in animals with an open circulatory system; diffusion is only fast enough for small organisms;

Q1.3 Blood can pass slowly though the region where gaseous exchange takes place; maximising the transfer of oxygen and carbon dioxide; and then be pumped vigorously round the rest of the body; enabling the organism to be very active;

Q1.4

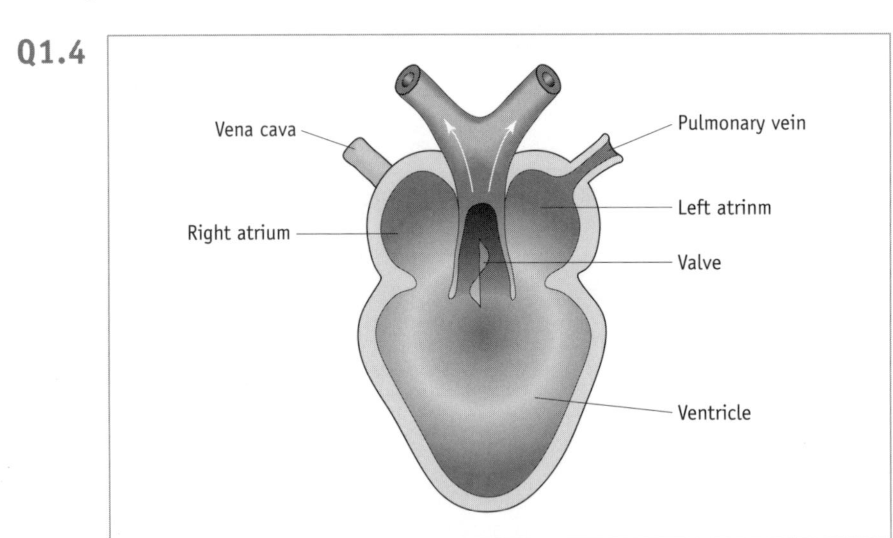

Q1.5 Some mixing of oxygenated and deoxygenates blood; in the ventricle;

Q1.6 Thick layer of mainly elastic fibres; and smooth muscle; to allow expansion of diameter of artery; surrounded by thick layer of mainly collagen fibres; to aid in contraction of diameter of artery;

Q1.7 **The blood pressure in the left ventricle falls below that in the aorta; leading to the closure of the semilunar valve between the right ventricle and the aorta; the blood pressure in the right ventricle falls below that in the pulmonary artery; leading to the closure of the semilunar valve between the left ventricle and the pulmonary artery;**

Q1.8 **Leads to a better quality picture; as a scanner can be moved more smoothly than can a person; more comfortable for the patient;**

Q1.9

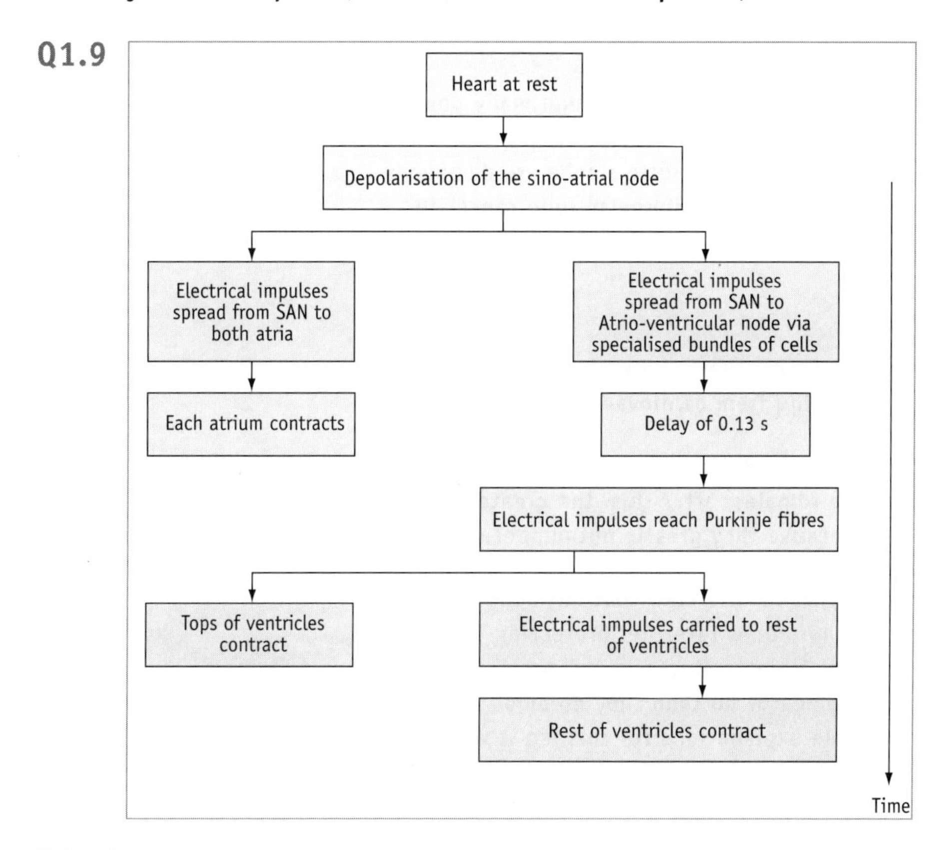

Q1.10 **Shaving and swabbing with alcohol removes hair and oils; improving electrical conduction; leading to a better quality ECG;**

Q1.11 **In a healthy person, exercise merely results in the ECG showing a greater frequency / rate of electrical activity; in an unhealthy person undertaking exercise the ECG may reveal abnormalities; which can aid in diagnosis;**

Q1.12 **Between the sino-atrial node and the atrio-ventricular node; or at the Purkinje fibres;**

Q1.13 **Very high heat rate; abnormal T wave;**

Q1.14 **Congenital; extreme exhaustion; extreme dehydration; certain illicit drug use;**

Q1.15 Recessive; phenotype only evident in individuals who have two copies of the allele in question;

Q1.16 a) A quarter / 0.25;
b) 1;
c) 0;

Q1.17 Require one group of people to smoke cigarettes; and another group of people; matched for such characteristics as age, gender, etc.; not to smoke; check individuals in each group over time to see if they develop lung cancer; analyse results statistically;

Q1.18 Not right / fair to force some people to smoke cigarettes; but the experiment might be acceptable if only volunteers were recruited;

Q1.19 Do the experiment on non-humans, e.g. dogs; though many people would think this just as unethical; carry out laboratory tests; for example on tissue cultures; to see if cultures exposed to cigarette smoke are damaged; but damage to tissue cultures might not necessarily mean lung cancer in whole living people;

Q1.20 It increases; greatly;

Q1.21 Not necessarily; it might be that behaviours early in life greatly affect one's subsequent chances of dying from cardiovascular diseases;

Q1.22 No; between the ages of 10 and 79 males are more likely to die from cardiovascular disease than are females; after this, the greater number of deaths among women is partly because they greatly outnumber men;

Q1.23 The data provide some support for the view that until the menopause a woman's reproductive hormones offer her protection from coronary heart disease in that deaths from cardiovascular diseases increase more steeply among women over the age of 50 than they do among men; quantify; but there are other possible explanations for this; so it would be premature to draw such a conclusion with any confidence;

Q1.24 Blood not pumped at normal rate to lungs; shortness of breath;

Q1.25 a) Approximately 480 000 calories;
b) Approximately 480 Calories;

Q1.26 Hydrogen and oxygen are always found in a 2:1 ratio / in the same ratio as in water; the ratio of carbon to H_2O varies;

Q1.27 30.1; moderately obese;

Q1.28 The risk of death from coronary heart disease steadily rises; from around 6.5 deaths per 1000 men a year with a serum cholesterol level of between 4.1 and 4.9 mmol per L; to around 11.5 deaths per 1000 men a year with a serum cholesterol level of between 8.1 and 9.6 mmol per L;

Q1.29 The fact that pre-menopausal women generally have higher HDL-LDL ratios than men would be expected to lead to their having lower rates of coronary heart disease;

Q1.30 Cheaper; basic foods that are high in vitamins contain much else that is enjoyable to eat; and of nutritional value; vitamin supplements can be valuable for certain people; but some do not provide all the vitamins that are needed;

Q1.31 Extremely difficult to tell; even if anxiety runs in families this might be because of the common environment family members experience; however, many human traits have a genetic basis; we know from dogs and other animals that a susceptibility to anxiety can be inherited;

Q1.32 We could learn to be more understanding of people who get very stressed; we could change things so that people were less likely to become stressed; medication provided on the National Health Service can help in some cases; other treatments are available;

Q1.33 Difficult to be certain; possibly by affecting the levels of particular chemicals in the brain; possibly simply by tiring a person out; possibly the benefit is psychological;

Q1.34 Antioxidants protect against free radical damage; free radicals are highly reactive; and can damage many cell components; they have been implicated in the development of heart disease and some other diseases;

Q1.35 30; detrimental;

Q1.36 They might decrease the permeability of the walls of the nephrons to water and ions; thus leading to less water and ions being retained in the body; and more lost in the urine;

Q1.37 A person may equate the statement 'your blood pressure is too high' with 'you are ill'; many people assume you automatically sign off sick when you are ill;

Q1.38 People who have had a heart attack may be more motivated to follow a strict diet; they probably started with higher blood cholesterol levels making it easier for them to reduce these; they may live in institutions where control over diet was greater;

Q1.39 This reduces both HDL and LDL cholesterol and the ratio may remain the same; lowering blood cholesterol levels without altering the HDL–LDL ratio may be ineffective in reducing risk;

Q1.40 Fruits; vegetables; cereals;

Q1.41 Most foods;

Answers to in-text questions for TOPIC 2

Q2.1 For A, SA = 6, Vol = 1, SA:Vol = 6; For B, SA = 24, Vol = 8, SA:Vol = 3; For C, SA = 96, Vol = 64, SA:Vol = 1.5;

Q2.2 (a) Its surface area increases by a factor of 4; (b) Its volume increases by a factor of 8; (c) Its surface area to volume ratio halves;

Q2.3 The surface area to volume ratio would continue to fall; the organism would not be able to exchange enough substances to survive;

Q2.4 Hippopotamus;

Q2.5 For D, SA = 34, Vol = 8, SA:Vol = 4.25; For E, SA = 28, Vol = 8, SA:Vol = 3.5;

Q2.6 Volumes are all the same, 8; but the more elongated the block the greater the surface area and thus the larger the surface area to volume ratio;

Q2.7 Tape worm;

Q2.8 Dessication / dehydration problems; surface also has protective function;

Q2.9 D;

Q2.10 Lungs; gut; kidneys; capillaries;

Q2.11 They are carried in the bloodstream;

Q2.12 Pathogenic microorganisms have time to multiply resulting in sickness;

Q2.13 Acid in the stomach kills the microorganisms;

Q2.14 The kinks in the fatty acids prevent them lying very close together; this creates more space in which the molecules can move;

Q2.15 a) Diffusion;
b) Active transport;
c) Active transport;
d) Facilitated diffusion / channel protein;
e) Active transport;
f) Osmosis;

Q2.16 Salt is normally reabsorbed from sweat using the CFTR channel; with CF this does not function; so the salt is not absorded making salty sweat;

Q2.17 It is a polymer; of nucleotides;

Q2.18 Answer in text immediately after question;;;

Q2.19 T A G G G A C T C C A G T C A;

Q2.20 A U G U A C C U A A G G C U A;

Q2.21 5;

Q2.22 UAC; AUG; GAU; UCC; GAU;

Q2.23 G T C A G T C C G;

Q2.24 a) The sixth base is G rather than T;
b) The second base is A instead of T;
c) The fifth and sixth bases are inverted;
d) Omission of the fourth base, A;
e) An additional base is added after the fifth base;

Q2.25 The 507th triplet now reads ATT; this codes for the same amino acid isoleucine;

Q2.26 In situations one and two both parents are probably carriers and there is a 1 in 4 chance that any child they have has CF; if the father in situation 3 is not a carrier then none of the woman's children will get the disease; they will all be carriers receiving a defective allele from her and a 'normal' allele from their father; if these children were to have children with a carrier there would be a 1 in 4 chance that each child would have CF;

Q2.27 Carbohydrases; lipases; proteases;

Q2.28 Success rates for heart and lung transplants are higher than for a lung transplant; partly because no cuts have to be made to the pulmonary circulation;

Q2.29 With a false positive unnecessary treatment could be given; with a false negative failure to start treatment could lead to poorer health in the long term;

Q2.30 The test will not be completely reliable (it is only about 80–85% sensitive); there should not be false positives but there will be false negatives where an individual has a CF mutation but does not have one of the mutations which the test can detect;

Q2.31 Yes; if the parents believe for religious or other reasons that abortion is wrong; if they consider that the risk of miscarriage is too high; if they want to have the baby even if it will have cystic fibrosis;

Q2.32 Have a child accepting the 1 in 4 risk of it inheriting CF; use artificial insemination by donor to avoid the risk; use *in vitro* fertilisation and have the embryo screened before implantation; use prenatal screening and decide whether to continue with the pregnancy if the fetus has the disease;